Left

A History of the Hemispheres

. . . .

Matt Gaidica

To my grandpas, Charles & John

Contents

Acknowledgments

Many special thanks are in order to an enormous amount of people who have made this project possible. My editor, Alicia Voss, whose edits and additions exceeded my expectations. Those who entrusted me with a pre-order — Tim Woods, Mike Lleshaj, Billy Lindeman, Laurie Johnson, Roberto Johnson, Brad Birdsall, Jay Zawrotny, Ryan Blagdurn, Joey DiStefano, Alison Hedke, Jill Davie, Jimi Varner, Jesse Masters, Eddie Schodowski, Minh Nguyen, Megan Michaux, Chris Gaidica and Courtney Zokas — your funds helped supply this project with printer ink, highlighters and determination. The University of Michigan, whose labs and libraries have become indispensable temples of learning. My comrades in Rome and Edinburgh who kept me company while researching overseas.

Finally, to my parents who have always nurtured curiosity and free thinking. For their never-ending support, encouragement, and of course, our rich and lively debates.

Preface

Not long ago, I resigned from a startup company in Silicon Valley. I had 2,300 miles of open road between that wonderful place in California and my hometown in Michigan to answer a very simple question: why is it so hard for people to get along? I was looking for a metaphor that could describe, in simple terms, exactly what is happening when two people cannot seem to reach a resolution. The strength of a team is not that everyone knows how to solve a problem—that is easy—but that they are all rallied to solve it the same way. If conflict itself could be put into simple, ubiquitous, and objective terms, perhaps everyone, including you and those around you, could achieve perpetual harmony. If teams had a proven rubric, format, and framework, maybe conflict would disappear.

When do we end up conceding to another? What compels us to do this? How can we make sure that, even in times of disagreement, progress is made?

Before long, I became convinced that the organ resting upon our necks houses the answer to all these questions. We usually don't think about the brain as something that weighs our opinions, thoughts, or actions before they stream upwards into our conscious minds, but the makeup of what we call *reality* is, in actuality, a gross emergence of billions of processes churning under the hood. There is a meteoric cascade of bloody battles and hearty handshakes occurring in our brains when we do anything from pick up a glass of water, to mark an answer on a test. We humans are a fantastic cacophony of electrochemistry, our heads being, as Charles Sherrington poetically declared, an "enchanted loom where millions of flashing shuttles weave a dissolving pattern." Unearthing the way in which the brain mediates became the Pythagorean theorem to my triangle—I felt as though once the basic machinery was gripped, a model would be exposed that could reveal universal truths about human nature.

Throughout this book, I will focus on what I believe to be the most important debate contained within our cranium: the dueling of our two cerebral hemispheres. Every memory we recall and every action we take is influenced by the very notion that there are two halves that must come together as a whole, and more importantly, that we have deemed one *left* and one *right*. When we consider the brain as two opposing entities, the connection between biology and sociology becomes outright blinding. The only difference is that the brain has been working towards order instead of chaos for many millions of years, and formal society is still but an infant in matters of solitude.

Have you ever wondered why you always sign your name and take notes with the same hand? In the course of writing this book, I have posed this simple question to countless people. Don't feel ashamed if you, too, just realized that you never really had a choice. It is not our symmetry that makes us special; it is indeed our asymmetry. One side of the brain commands language, and the other explores faces and landscapes, so why don't we feel lopsided as those particular talents are thrust into action? Instead, what we *feel* is unified. Our brain is harboring the precious secrets of a cerebral peace treaty in every moment.

The spark for this book resembles nothing of the flames it created, although I do hope a seed has been planted which can make use of any light I may shed. From here onwards we are talking about the brain. This is a journey to understanding the nature of our two sides that spans across many topics, from medical history to ancient culture, from religion to the beginnings of the cosmos itself. Since that long car ride across a vast America, I have been on a magnificent expedition to the center of ourselves, and it is my pleasure to share it here with you.

MATT GAIDICA
UNIVERSITY OF MICHIGAN

CHAPTER ONE

Localization

Have we, in fact, two brains as we have two
eyes, two lungs, two kidneys? Or have we one brain
as we have one body, built up of two similar halves?
— HENRY MAUDSLEY

The concept of progress is a deep prejudice with
an ancient pedigree.
— J. B. BURY

I would rather be the discoverer of one fact in
science than have a fortune bestowed upon me.
— ROBERT KNOX

* * *

EDINBURGH, LATE NOVEMBER, 1827—It was a brisk
evening on West Port Street, one of the most central pathways through Edinburgh's lower-class districts. By day, it
was a bustling marketplace, alive with the rapture of peas-
ants, bakers, and broom-makers. By night, it transformed
into a stumbling grounds for drunkards and the type of
figure hailing from the lower echelons seeking pleasures
to make them forget of their pains. Orwell's proles, Huxley's epsilons. Two men were headed eastward, fitting

comfortably in the shadows, silhouetted by the yellow glow of gas-fueled lamps, destined for a place called Surgeons Square, which happened to be the epicenter of modern medical science.

Witness to their every move was Edinburgh Castle, sitting high above them in the clouds and mist. This twelfth century hill fort perched atop an extinct volcano, forged from sandstone that swept from high walls to a large circular chemise, undeviating as it lowered into the mutely colored mountains below. On the men's backs and within their begrimed hands was a large withered rucksack, apparently heavy, and quite awkward to carry. As they continued on their way, their wood-heeled shoes skipped across sharp gravel and uneven brick pavement with a piercing echo that ricocheted off of the densely packed, six-story tenements, permeating like a spirit into the wynds and closes that hid to each side of the street.

It was less than a mile before they approached a three-story estate—No. 10 Surgeons Square—contrastingly salubrious from the heart of Old Town from which they had trudged. The stouter of the two characters, standing round-faced and proper in his blue Frock coat and Bedford crop hairdo, revealed himself in the doorway, only to be sharply interrupted by a young voice from inside with the ensuing exchange.

"Were you looking for anyone?" the young man said, as he peered into the dour-looking face of the stranger.
"Umph! Are you Dr. Knox?"
"No, but I am one of his students," was the reply of the young man, who was now nearly well satisfied as to the intention of the stranger whom he had accosted.

2

> *"And sure," observed the latter, "I'm not far*
> *wrong thin, after all."*
> *"And I may suit your purpose as well,*
> *perhaps."*
> *"Perhaps," answered the strange man,*
> *"perhaps you may, sir."*
> *"Well," said our friend, the young student,*
> *"don't be at all afraid to speak out. Tell me your*
> *business, although I have myself an idea as to what it*
> *may be. Have you got 'The Thing?'"*
> GEORGE MAC GREGOR, THE HISTORY OF
> BURKE AND HARE

Words issued with a faint mist from the men's lips on this cold winter's night as they conversed about The Thing. It was not the frigid air raising goosebumps though; it was the fact that both parties were amidst the barter of a dead body. The contents of the lumpy burlap that had just been shuffling through the streets was a human being, one no longer able to taste the chill in the air, to think a single thought or hear a single word, his lips and fingertips blued by ischemia, lifeless among these three racketeers.

* * *

The Scottish Enlightenment was well underway, making the capital city of Edinburgh home to the most celebrated intelligentsia of all subjects. The constellation of human achievement included Adam Smith, David Hume, and Sir Walter Scott, who would form the legacy of what was later called a Modern Athens. This was the epoch that broadened studies in divinity and humanities into natural and moral philosophy, political economy, sociology, linguistics, and history, among many others. Burgeoning under

3

their feet was one story that begs to be retold, a subject that would triumph in the minds of the inquisitive and taint the souls of the men who ran the underground markets: anatomy.

Dr. Robert Knox—the homeowner being asked for by the crooks from Old Town—was a rising star in the growing anatomy scene, having made a name for himself while serving in the British Army. He pioneered the studies of the ciliary muscle in the eye, which is responsible for controlling the small contractions that assist us in shifting focus. He was a man very much drunk off the spirit of science. Well before Charles Darwin, he had spent time exploring the ethnologies of wild Africa, examining the apparent fish gills in the human embryo which only later transformed into lungs, and had expended great effort in trying to explain the commonalities among all life forms.

Knox's depth was accompanied by a genius wit for which he was well-known in his lectures. It was said that five minutes of Knox were equal to any other man's half hour and that he became a guide, philosopher, and friend to every worthy student. His dedication to his growing number of pupils went so far as to guarantee that every one of them had the privilege of seeing a human body dissected. Marie Bichat, the French anatomist of a century before, justified and promoted this type of investigation, which was beginning to raise eyebrows, when he said, "You may take notes for twenty years from morning to night at the bedside of the sick and all will be to you only a confusion of symptoms," and without opening the body, all you might turn out with was "a train of incoherent phenomena." For although the enlightenment was notoriously liberal in its acceptance of new science, the act of "peeking inside" was not one quickly embraced. Knox

4

was in good company, though, as Louis Pasteur in France, Robert Koch in Prussia, and Joseph Lister in the town of Edinburgh itself were putting germ theory into practice. The public was beginning to accept doctors and surgeons as the new healers, preferred in growing numbers to the long-instated barber surgeons who could drain and shave you all in one sitting.

It was not just Knox, but the whole of Edinburgh's medical community, that relied on fresh cadavers for research, a good sometimes provided by shady characters from the West Port district. While the importance of anatomy was slowly being warmed up to, it had not yet won out over the sacredness that was bestowed upon a breathless body. At the time, the laws permitted only the bodies of executed criminals to be used by anatomists in dissection—these malefactors were soulless, anyway! Polarizing articles appeared weighing the popular opinions on the matter, such as using the poor or those in public debt for research, resulting in a heated debate among scientists, politicians, and diviners. One such article entitled, "The Use of the Dead to the Living," by Thomas Southwood Smith motioned for a change in policy, ultimately targeting the poorest of the town:

> *These persons are pensioners upon the public bounty: they owe the public a debt: they have been supported by the public during life: if, therefore, after death they can be made useful to the public, it is a prejudice, not a reason—it is an act of injustice, not the observance of a duty, which would prevent them from becoming so.*
> THOMAS SOUTHWOOD SMITH

Despite the sharp rise in Edinburgh's population at the dawn of the nineteenth century—to well over 200,000 people—there were still not enough bodies coming from delinquents to satisfy the thirst of the anatomists.

The chronic shortage of bodies gave rise to *The Resurrectionist Movement*. Of the many rackets that littered the city, the *violation of sepulchers*, or grave digging (usually accompanied with an intent to sell the body for dismemberment), was regarded as one of the most foul. It often forced family members to stand guard of their kin's gravesite, and in more extreme cases, an iron cage called a mortsafe was installed over the gravesite to hinder would-be thieves.

At this time, the Scottish had an almost superstitious reverence for the dead, and their belief in resurrection was extremely material. It was thought impossible by many that when, as George Mac Gregor states it, "last trump should sound," that the dead could rise if the bodies were cut up in dissection.

Even Shakespeare was concerned about the sanctity of his rest after death, as shown by the epitaph which adorns his grave in Westminster Abbey:

> *Good friend for Jesus sake forbeare,*
> *To dig the dust enclosed here.*
> *Blessed be the man that spares these stones,*
> *And cursed be he that moves my bones.*
> WILLIAM SHAKESPEARE

The moral, ethical and spiritual ramifications are hardly a new subject to the human species. Although keen to the methods, Egyptians opened bodies only in the process of mummification but never to pursue natural inquisitiveness. In the pages of the Quran, Muhammad explicitly forbids this practice. The ancient intellectual metropolis of

Alexandria was the first to produce systematic dissections in the pursuit of greater knowledge by Claude Galen; however, his anatomy sessions were limited mostly to animals, and were few and far between at that.

It was not the innate human repulsion to seeing blood and internal tissues that hindered the advance of human dissection; rather, the issue delves much deeper into philosophies to which many cultures prescribe, considerations of mind-body dualities, and religious sentiments which seem to be ingrained in us all.

However, in 1829 Edinburgh, one thing was quite clear: reigning physicians and even students of polite society had very few qualms about buying bodies delivered from ominous characters trudging in from across town. The body bartered in Surgeons Square had been a result of more perverse circumstances than simply being dug up, because the salesmen were not even resurrectionists. The sacked man was a lodger that died after failing to pay his monthly pension to the portly, Frocked-coat landlord who was now privy to the diabolical scheme to reconcile the payment by selling the old man's body. The final negotiation between Dr. Knox's associate and the two men ended at £7 10s—to later be split in half—which was a small fortune for a hawker and a cobbler. Little did the duo know, this encounter would be the spark to a series of events much more gruesome than any transaction that could arise from natural causes. It was the beginning of an infamous chronologue in the history of anatomy, being forever a source of macabre material for the darkest literary genres.

These newly entitled entrepreneurs of humble beginnings were effectively lured into the ghastly business of providing fresh cadavers by any means possible. They had circumvented the muddy and risky digging process altogether, and with that, provided a delightfully fresh product, free from the typical maggots and earthworms which anatomists could hardly stomach. It was "getting that high price," as one of them said, that authorized their murder-for-hire scheme. At the culmination of their exploits, they had murdered at least sixteen innocent people, nine of the acts committed at the very residence at which the unfortunate pensioner died, forever branding the events as *The West Port Murders*, and the twosome by their surnames as *Burke & Hare*.

William Burke and William Hare

Sir Walter Scott piped in on the situation, saying, "Our Irish importation have made a great discovery of Oeconomicks, namely, that a wretch who is not worth a farthing while alive, becomes a valuable article when knockd on the head & carried to an anatomist...." This was a clear blow to the immigrants who found permanent residence in Edinburgh, and highlighted the sinister game of supply and demand that Burke & Hare stumbled into.

The final victim in the serial string of smotherings drew some attention from the public and led the police directly to Surgeons Square, where the body was found in the possession of Dr. Knox. He was able to forgo prosecution by playing dumb, asking why he should be the one to care about how the body came to him. Although it was nearly certain that a man of such distinction knew well of the procurement methods, dead was dead to the anatomist, and it was presumably not in their job description to care much about how it came to be. As the murder cases unfolded, Knox's reputation became repugnant and led to his expulsion from the city. He would later find work in London, but was never able to rise to the stature he had once held in Edinburgh. Little remembered are his seminal works, and rather instead this scurrilous jingle:

> Doon the close and up the stair
> But and ben wi' Burke and Hare
> Burke the butcher
> Hare the thief
> And Knox the boy that buys the beef.
> SURGEONS' HALL MUSEUMS, EDINBURGH

Burke the butcher, the stouter of the criminal duo who initiated the conversation with the student, was prosecuted for the atrocities after his partner in crime, Hare, became an informant to the King in the investigation. Scottish aristocrat and literary gossip Charles Kirkpatrick Sharpe prefaced the leading account of Burke's trial by saying it was "dreadful proof of the deep depravity of human nature." Burke was hung before the public on January 28, 1829, and in the ultimate display of irony, dissected in public by the Royal College, an institution that Knox vehemently loathed.

The city would go on to pass the Anatomy Act of 1832, which put an end to the so-called "anonymous subject," requiring that anatomists record the cause of death of each individual. Most importantly, it gave licensed anatomists access to the bodies of not only murderers, but unclaimed workers and prisoners, provided that after the dissection they pay for a proper burial. Hundreds, if not thousands, of bodies bought and dissected before the Act remain "unknowns." They were never tracked, leaving no legacy for their families, and their body parts were casually dispersed into the unvisited corners of Surgeons Square, gruesomely forming the foundation upon which anatomy now stands.

* * *

The student who conferred the sale of the poor pensioner in Edinburgh would later become a premier surgeon of the country, adding substance to a looming paradox that beseeched the medical community of the day. The famous proverb asks if 'the ends justify the means'—if killing one is justified by saving thousands. This was certainly one of Burke's more utilitarian alibis during his trial. To what extent are we as humans able to judge, contemplate, or indemnify others for acts of atrocity? Human history is deeply entangled with cleansing and genocides pursued in the name of a higher cause or power. The Sioux at Wounded Knee Creek, the Jews across Europe, the Maya of Yucatán, the Dzungar in China, the Tutsis in Rwanda—all are stark reminders that despite our benevolent self-reflection, we are a brutal bunch.

No matter the possibility of positive contributions by means of William Burke to better society through his supply chain of specimens, he was quite plainly a murderer and responsible for a whirlwind of abhorrence in the public mind. He was cast an animal and a moral outlier, one of the first serial killers on public record, which made Burke himself a unique subject for further investigation.

Elevated just one block off the street that Burke & Hare used to deliver their bodies, and only a matter of days after the infamous trials came to a close, a group of distinguished thinkers were in fierce debate over why and how Burke was such a monster. Being passed around were two pieces of paper that served as all the evidence these men would need in their explanation. The first was a side-profile drawing of Burke's shaved head made during his autopsy. The second piece of paper was titled "Measurements," and began with the following data set:

Spine to Individuality, ... 8
Concentrativeness to Comparison, ... 7 3-8
Ear to Occipital Spine, ... 4 4-8
— to Individuality, ... 5 2-8
— to Firmness, ... 5 7-8
Destructiveness to Destructiveness, ... 6 3-8
Secretiveness to Secretiveness, ... 6 2-8
Ideality to Ideality, ... 4 6-8
Constructiveness to Constructiveness, ... 5 2-8
Cautiousness to Cautiousness, ... 5 6-8
THE PHRENOLOGICAL JOURNAL

To any modern physician, psychiatrist, or medical doctor, these documents would bear no value in formulating a diagnosis for Burke. But this was not the case for the Edinburgh Phrenological Society, which was, at the time, thirty years into a growing discipline of science which suggested that there was an undeniable relationship between the cranial structure and propensities of an individual. These men believed that through a detailed analysis of Burke's head and the measurements of his skull—marked on their papers as a distance of inches followed by tenths of inches from locale to locale—his true murderous nature would be revealed. And this was not a system used only for murderers like Burke. It was a holistic, systematic discipline developed by the Society that could explain how any man or woman had fallen from grace based on the shape of his or her head.

The system of phrenology was the product of that curious type of genius that lies right next to absurdity, beginning with a playful mind pondering the mechanics of the world. You could say that Franz Joseph Gall was this type of prototypical dreamer. Starting early in his childhood, he was wholly interested in the variation in talents of others around him. He noticed that everyone had a particular aptitude, or deftness, that distinguished them. This insight did not come as a spark, but rather a dithering flame that persisted throughout his youth, as he observed that the abilities of penmanship, grammar, and mathematics were unevenly distributed among his peers. The true test of universality came during a move, after which he noticed that individuals at his new school who had a talent for memorizing things had the same "prominent eyes" of those who had possessed this ability at his old school. A revelation—an epiphany!—came unto young Gall! His hypothesis stuck with him into University, and he even convinced some of his classmates that it was a phenomenon which was far beyond mere circumstance. Little did he know then that this idea would consume and define his professional career. If memory was an attribute that could be ascertained from a physical quality of the human form, perhaps, he thought, this may be the case for all intellectual powers. He would look to the place where the body coalesces into a person, where true character is formed, upwards into the bones that form the cavity for the brain.

* * *

The march of history provides a variety of blunders when attempting to seat moral sentiments on the geography of the body. Egypt is credited with the first official records of brain anatomy, outlined in an antiquated surgical treatise dating back nearly 5,000 years. Although the ancient Egyptians appeared to have known some details about the human brain in these cerebral manifests, their postmortem embalming procedures called for the removal of the brain using a hammer, chisel, hook, and spoon, which presumes that they placed little value on its accompanying them into the afterlife and the *Fields of Yalu*.

One of the first people to place the soul and thought processes in the brain was Pythagoras (around 550 B.C.), the same philosopher whose geometrical theorems are familiar to us all. Less than one hundred years later, Hippocrates, now deemed the "Father of Western Medicine," believed the brain was the source of "joys, delights, laughter, and sports, and sorrows, griefs, despondency, and lamentations," and by this, "we acquire wisdom and knowledge, and see and hear and know what are foul and what are fair, what are bad and what are good, what are sweet and what are unsavory." To Hippocrates, the brain was almighty and all-powerful. His metaphorical and poetic rendition of our material selves consequently began what became known as *humanism*. The next generation ushered in Plato, who prescribed to the Hippocratic view, though his student Aristotle would choose to follow (and fall into) the cardiocentric hypothesis: an opinion that the brain and lungs were simply cooling mechanisms for an all-powerful furnacing heart. The next time you sign a note with a heart to your loved one, just remember that you are reinforcing a very old theory about where love emanates from. Efforts to seat the mind finally begin to cool with Galen, whom I introduced earlier as a dissectionist. Although he was still alluding to the use of pneuma and humors as the body's modus operandi, he began suggesting the specific locales of the body which would be appropriate for the mind. It was becoming more true with each generation that the *self* was most specifically a product of our bulging brains.

As Dr. Gall continued to garner support for a brain-centered mind into the nineteenth century, his contributions tended to be those which related to his new theory of head measurements. To his misfortune, he would die in Paris just six months before the Phrenological Society began their investigations on the most prime specimen, Burke the butcher.

The Society's founder and supreme evangelist, George Combe, questioned what is truly inside the skull, "an airy dome; a richly furnished mansion; one apartment, or many." However, phrenology's suggestions were clear—there is a true maisonette of the mind—totaling twenty-seven organs within the cranium, each with clear-cut distinctions in form and function. There was not only a division of labor, but it was hierarchical, with areas of unequal worth ranging from the lower animal propensities, to higher centers of intellect. As Combe reviewed the case of Burke, he triumphantly proclaimed that "phrenology is the only science of the mind which contains elements and principles capable of accounting for such character as that before us," and that it is a science of "unbending truth."

Combe's official review of the William Burke case appeared in the Edinburgh Phrenological Journal only nine days after Burke's execution, citing measurements from Burke's head taken in life and after death.

The posterior and middle lobe, home to the organs of Destructiveness, Secretiveness, and Acquisitiveness, were very large indeed; Destructiveness had to it a distinct swell.

Thus, Burke's animalistic tendencies and feeble moral sentiments were fully accounted for. His Philoprogenitiveness was of considerable size, which was reconciled by earlier stories of him regularly passing candies out to the children of his neighborhood. His Love of Approbation was also well developed, expected from a man known to have exhibited love for his wife, priest, and parents.

The exposition couldn't go without making note of a "withering scowl which defies all description," that was observed by a prison escort moving Burke from his cell to the gallows, a revealing of his most evil propensities. Burke was a child charmer by day, and a ruthless strangler by night. He was described by Combe as being able to front a "calm exterior," which made him ever more dangerous to society.

Burke came to be the quintessential double-faced caricature used as a source by many novellas, including Dr. Jekyll and Mr. Hyde, not coincidentally produced by one of Edinburgh's finest, Robert Louis Stevenson. But was he really a serial killer built in the womb, or was he simply shaped by his circumstances? Were the pressures of Destructiveness so great throughout Burke's life that it modified his cranial structure? As evil propensities took over in strength, could it be that the ones for good atrophied? Was it really possible to distinguish the crania of murderers from others?

*　*　*

In building theories, all scientists share a similar constraint to their work—they must start somewhere. The list of references found at the end of any academic paper or book is usually representative of that *somewhere* for the author, with their work being a creative mixture of those ideas, taking the topic to a new and wonderful place. In this way, it is the experiments of the past which provide a basis for discoveries of the future. The philosophical metaphor of *Neurath's Boat* likens this process to rebuilding a boat in the middle of the sea, remembering that you must always have a place to stand. These necessary foundations of science are as known as *a priori*, or "from the earlier," and become preconditions to inquiry. There is no way around it, since progress can only come through the small steps of many, though it presents the opportunity for disciplines like phrenology, geocentrism, creationism, and atomism to flourish.

The a priori knowledge from the days of phrenology in Edinburgh were sufficiently passed into the life of a prominent surgeon in Paris during the latter half of the nineteenth century. Paul Broca is one of the figures that shaped modern brain surgery, neuropsychology, and neurolinguistics, with his accomplishments extending far beyond medicine. He was known as a man of both warmth and brilliance. It was, however, no secret that much of his work was centered on the question of intelligence of different races and sexes, and much of his investigation leveraged the burgeoning tools founded by the phrenology and craniology movements.

Broca was heavily influenced by Samuel Morton, who was not coincidentally a University of Edinburgh graduate. Morton was known for his collection of more than one thousand skulls in his house that many referred to as "The American Golgotha," and which he displayed not as relics, but rather as objects of study. Morton was set on empirically ranking races, with the most notorious attempt in his first volume *Crania Americana*, published in 1839. In this he used detailed lithographs, measurements, and recorded observations to make a case against the indigenous people of North and South America.

Paul Broca

Morton often cites the work of the phrenologists in his discourse, taking the literature as scientific fact rather than theory. Broca was largely in defense of Morton's ideas, once noting that the study of the brains of human races would lose most of its interest and utility to any anthropologist if there were not definite variations to be observed. The art of craniometry, or anthropometry, was sourly driven by the political, and perhaps moral, need to prove that Europeans were superior during the nineteenth century. Broca and Morton were a new breed of phrenologists, yet they were more concerned about measuring man on scales of intelligence and overall capacity than citing centers that correspond to the endless list of phrenological propensities.

Although apparently a victim to some a priori convictions, Broca was a keen scientist and investigator, and ironically responsible for his protege's professional motto, "*J'ai horreur des systèmes et surtout des systèmes a priori*," or, "I abhor systems, especially a priori systems." He can no more be condemned by his thoughts or actions than can anyone else when taken in historical context, and his reputation deserves some defense. In *Broca's Brain* by Carl Sagan, a diligent effort is made to relieve the surgeon from his prejudices, and he is marveled as a progressive man thinly bound to the ideals of the world of which he was a product. Sagan writes vividly of his tour to Musée de l'Homme, or "Museum of Man," in Paris. As he walked through the innards of the museum, areas off-limits to the public, past the "warren of dark, musty rooms, ranging from cubicles to rotundas," overflowing with collections of antelope bones from the Paleolithic, throwing spears from Oceania, skin drums and ceremonial masks, he came to a most-macabre collection. It was a collection started by Broca himself, of which he would eventually become a part. The glass jars that filled the room were of all shapes and sizes, each containing some type of specimen submerged in formalin. The title of Sagan's book is a result of peering into one of these jars labeled, P. Broca.

Were grand memories still locked inside Broca's embalmed brain of dinners with Victor Hugo, and strolls along the Quai Voltaire and the Pont Royal?

For Sagan, Paul Broca was worth writing about because of his empirical intensity and his fervent studies of "embryos and apes, and people of all races, measuring like mad in an effort to understand the nature of a human being." It was this measuring like mad that set Broca apart from others. The discoveries that would forever cement Broca's name in text-books did not come from his studies on race or sex, as he himself acknowledged that these were a rather selfish pursuit.

> *It is the natural tendency of men, even among*
> *those most free of prejudice, to attach an idea of*
> *superiority to the dominant characteristics of*
> *their races*
> PAUL BROCA

His most seminal work came from autopsies he performed on patients that suffered from a specific speech impairment; they could understand language, but when they spoke, they never managed to say more than a couple syllables, of which none had meaning. The oratory apparatus was intact, they had normal intelligence, and could even understand written words, so it was a specific defect in articulation that Broca cited as the problem. The most famous of his patients was "Tan Tan," named this because the only thing he could say was his name, "Tan," and it always came out in bursts of two.

Broca eventually accumulated overwhelming evidence on the condition which he first called *aphémie*, coming to agreement with others who had been convinced that all faculties of higher intelligence resided in the frontal lobes. What was most striking though, was Broca's deviation from the long held law of "organic duality and functional unity." This stated that if something appears symmetrical, then it should act like it too! This is however, not what Broca found, and out of all the case studies under his review, language deficits were almost completely found with those who had damage to their left hemisphere. Not only did these findings suggest that the long-disputed idea of functional localization was true, but it said that these functions had an asymmetric nature. Not even the quacks who came up with phrenology thought that this could be the case, as their faculties existed in pairs between each hemisphere.

> *This proposition is no doubt strange, but however perplexing it may be for physiology, it must be accepted if subsequent findings continue to indicate the same viewpoint.*
> PAUL BROCA

By 1865, it was hard to argue with Broca's findings, and his triumphant proclamation, *Nous parlons avec l'hémisphère gauche*, or in English, "we speak with the left hemisphere," but many had their reservations. George Combe's exposé on phrenology, *The Constitution of Man*, was a best-seller, however it was slowly shifting into a pseudoscience. Psychics, mystics, and gypsies began offering "phrenological readings" that turned out to be no more telling than a modern palm reading or horoscope. There were also those who were adamant about interjecting religious sentiments into scientific fact, declaring that localization undermined the unity of the soul, and went as far as saying it opposed the very existence of God. If those things which made us human—such as speech—could be so easily removed from our repertoire, the Church was no longer the transcendent authority on man's place in nature. Not surprisingly, this was a big issue for a nineteenth century conservative French Monarchy that had a strong alliance with the Catholic Church. The freethinkers and libertarians latched onto this movement not only because it opposed the crown, but because anatomy had provided the ground for a full fledged revolution into an era of materialism.

The discovery of a language locale was altogether a *cause célèbre* for those fighting the stymieing ideologies of a perfectly symmetrical existence that had reigned for so long, but the question still remained, why might it be found on the left?

CHAPTER TWO

Evolution

Darwinian Man, though well behaved, At best
is only a monkey shaved.
— William S. Gilbert

Our reverence for the nobility of manhood will
not be lessened by the knowledge that Man is, in
substance and structure, one with the brutes.
— T.H. Huxley

Mankind is poised midway between the gods
and the beasts.
— Plotinus

*   *   *

It is worth a brief intermission to treat the concept of a cosmic calendar, giving context and prelude to the beautiful story of our formation. This metaphor—of crushing the entire history of our universe into a single year's calendar—certainly has the potential to radically alter your perception of our existence.

The Planck Cosmology Probe just recently concluded a four-year mission in outer space, collecting data on cosmic microwave background radiation and giving a more accurate estimation as to the age of our universe—now taken to be approximately 13.8 billion years old. This measurement is based on the Big Bang Theory, which suggests that long ago a single point, both infinitely small and dense, erupted and sent particles screaming outwards into empty space in all directions. For nearly four hundred thousand years, these particles expanded with such ferocity and chaos that photons couldn't escape clashing with the other loose elementary particles, thereby trapping them in an outer layer of plasmic dust. The energy of this expanding sphere slowly dissipated. Cooling ensued and the lawlessness of nature was overthrown by order as atoms began to form. No longer impeded, the photons finally escaped from the outskirts—the border of a now-expansive universe—thereby forming what can only be described as an omnidirectional, light-transmitting fog or "afterglow." This is the isotropic radiation measured by the Planck satellite, which is used in the reigning astrological models for dating the universe. It forms the perimeter of, quite literally, everything.

Dating is important because it determines how events are distributed along the cosmic calendar, forming a rich contextual vantage point. January first is marked as the day of the Big Bang. It is worth asking, what about the day before? We don't know, and it is doubtful if we ever will. A defining feature of any theory involving a singularity is that it maintains no artifacts of previous time; it is much too turbulent an environment for information to remain ordered and sequenced. Those who start afresh after the ball drops on New Year's Eve are quite familiar with this notion. It is a forgetting of everything in the past, and it begs to be asked if this human tradition has its roots in the behavior of the cosmos itself. Was it a previous universe that formed atoms like ours? Were there humans like us? *Were* the humans *us*? These are secrets to which the cosmos provides no answer, and the only solace you are to find is surely in a book of fiction.

Following the Big Bang, the universe continued its expansion through the spring months of the cosmic calendar, and our galaxy, The Milky Way, settled into its spiral by May eleventh. Summer quaintly passed by, largely uneventful, but beginning in September, our suburb of the galaxy was forged, starting first with the sun, and followed by the orbiting planets that filled in shortly thereafter. By September twenty-first, the first single-cell life forms emerged on Earth, DNA and all. It took the next two cosmic months for these single cells to work cooperatively as multi-cell organisms, making their debut on December first, nearly one billion years ago. Almost every organism on Earth today relies on these cells and the processes that they themselves perfected. They are the instigators of what would be a snowball effect of monumental proportions in the coming days.

Between December twentieth and cosmic-Christmas, insects, fish, plants, and reptiles spawned. The day after Christmas came mammals—small rat-like ones—that lived among the dinosaurs, in lands of great jungles, barren plains, expansive bodies of water, and with mountains still in violent ascent. The next day, December twenty-seventh, one hundred and fifty million years ago, the first birds took flight. December thirtieth marked both the death of the dinosaurs and the birth of primates. The Earth was no longer in a true Pangean form, though North America and Europe were still tightly embraced.

The last cosmic day arrived on December thirty-first, nearly 2.5 million years ago, and it wasn't until 10:24 PM that the most primitive humans graced planet Earth (you will meet one of them very soon). Eight minutes before the new year, modern humans lived in Africa, creating fire, communicating with ease, developing culture, and establishing some sense of self-identity through jewelry, body paint, and clothing. The first writing came into play fifty-five thousand years ago, a short thirteen seconds prior to the year's end. In the last ten seconds, astronomy was born out of Egypt, the lives of Buddha and Confucius came and went, Euclid invented geometry, Archimedes invented physics, *The Iliad* and *Metaphysics* were written, and Alexander the Great conquered and ceased. In the last two seconds the Renaissance dawned, and the final second encompassed that which is tied to our closest ancestors, and inevitably, ourselves. We are simply a speck in a corner of a universe and our existence encompasses but a fraction of a second on the cosmic calendar.

We have a deep seated, blissful ignorance to this overwhelming scale as we approach the great mysteries around us. Ever since we rose to our feet the horizon has teased us with a world more than the ground below. The expansion of space, complexity of the brain, and multifarious nature of social discourse encompass problems that are wildly disproportionate to our humble stature, but we have somehow learned—or shall we say, evolved—to overcome them.

* * *

She was a quiet girl from West Africa, born late in the summer of 1965 and soon after adopted by two loving parents in the United States. She was taught sign language, because any spoken conversation immediately induced a profound sensory overload, making verbal comprehension impossible for her. This minor setback did not prevent her from enjoying a pleasant childhood filled with afternoon tea parties, painting, and perusing shoe magazines. One of the games her parents played with her was a simple guessing game that may ring with your childhood nostalgia. Using photos of all types of animals, the little girl would have to split them into "humans" and "not humans" — even *Sesame Street* embraced this familiar activity on a weekly basis to the tune of "One of these things just doesn't belong!" It was fairly straightforward and easy for her, and she would always place pictures of herself and her parents in the "human" pile, while placing the ducks, lizards, and fish into the other one. Her name was Washoe, and if you hadn't already guessed that there was something else unique about her, she was a chimpanzee.

Although privileged by many standards, her seclusion and "homeschooling" secured for her a sheltered life, with only rare glimpses of the outside world, which frightened her beyond expression. When she was first introduced to other chimps at the age of five, Washoe told her parents through sign language that they were "black bugs." These monsters were just as despicably different from her as the Arawaks were from Christopher Columbus when he first set foot on the shores of Haiti. Washoe taught us a valuable lesson that exists even on the most primitive level: that the process of classifying species is prone to bias.

Classifying animals has always presented challenges. Today we are able to use evolutionary theory to guide this process, taking into account not only the appearance of a living organism, but their biological attributes too. Despite the most recent developments in DNA sequencing, which subtly and beautifully reveal the interconnectedness we maintain with Washoe, many people remain unconvinced. With nearly 40% of its citizens denying it as a valid scientific claim, America is one of the least accepting nations of evolution. However, as witnessed in the on-going, and high profile Texas school board debates, this is not strictly a matter of paleontology, biology, or archeology — as it should be—but one that resonates much deeper to some human beings. Should non-scientists be dictating what is included in a biology textbook?

There is no reason that evolution should remove the beauty from our world, or the souls from ourselves. The fossil record displays a magnificent representation of transitions and morphologies that, when viewed in its breadth, invoke a feeling of transcendence. A guinea pig can sit in the palm of your hands, yet has an almost identical bone structure to an eighty pound capybara. In his thousand-page treatise, D'Arcy Thompson shows how the skeletons of different horses, fish, crustaceans, and dinosaurs can all be morphed into each other under strict mathematical transformations. In other words, we can model the pressures of natural selection in terms of the geometry embedded in our bone structure. It is all too curious that nearly every mammal has five (and not six) fingers, and two (not three) arms and legs. The embryo of every animal begins nearly indistinguishable from others, outfitted with both gills and a tail. All of these differences among individual species, meticulously superimposed through the layers of the Paleozoic, Mesozoic, and Cenozoic eras, make for one very convincing case for an evolutionary process.

This theory remains an evidential beast, and, in the rational world of science, its acceptance must be predicated on predictions, then observations, and finally applications. It would be against the statutes of science to accept it on any other terms. Let's not forget that the word "theory" should not stifle the progress and applications of evolution. It is Einstein's theory of relativity which is hardcoded into all Earth-orbiting satellites that help us navigate on the roads and in the air. Cancer diagnoses and treatments are grounded in cell theory, electronics in quantum theory, and the combustion of gasoline in the theory of thermodynamics.

So whether you like it served punctuated or Darwinian, as the method employed by God or for evidence against Genesis, we shall move onwards with some acceptance and adherence to the notion that we can look at fossils of other creatures and elucidate something about ourselves. We start with the most disgruntling aspect of this reality: *we are all just a bunch of animals.* We must look to our differences in pulchritude, intelligence, and ferociousness in order to truly understand our place and purpose.

The stage was set early for a great *scala naturae*, or "chain of beings," by Aristotle, but it would take nearly two millennia before Carl Linnaeus introduced his "natural classification" in the 1700s which put life into a well-defined hierarchy. Although it wasn't suggestive of evolution (in fact, species were strictly said to be a result of Creation), it was meant to show that there were fundamental and underlying rules and patterns in nature. Linnaeus is responsible for standardizing the Latin naming conventions of organisms down the taxonomic rank (e.g., *Homo sapiens*), which range from and to: domain, kingdom, phylum, class, order, family, genus, and species. The cover of his most famous work, *Systema Naturae*, displayed his motto, *Deus creavit, Linnaierus disposuit,* or, "God created, Linnaeus organized."

The scientist in Linnaeus fell victim to the saint. He was so utterly convinced that each and every organism he studied was hand built by a creator, that he missed what was so obvious to Charles Darwin. The publishing of the *The Origin of Species* marks the moment that a true scientific theory would flourish and describe the source of all natural variation that was honorably pursued in past generations. Everything looks so similar because it is! It takes Darwin some time to be brutally frank with his readers, concluding his second most famous work, *The Descent of Man*, with this beautiful statement about his findings:

> *I have given the evidence to the best of my ability; and we must acknowledge, as it seems to me, that man with all his noble qualities, with sympathy which feels for the most debased, with benevolence which extends not only to other men but to the humblest living creature, with his godlike intellect which is penetrated into the movements and constitution of the solar system — with all these exalted powers — man still bears in his bodily frame the indelible stamp of his lowly origin.*
>
> CHARLES DARWIN

* * *

Oddly enough, Darwin began his education in Edinburgh as a medical student in 1825, giving the distinct possibility of his having brushed shoulders with Burke & Hare on his way to class. Shortly after recognizing he was more a naturalist, and less a surgeon, he left for Cambridge, where his studies of animals were firmly pitted against religious doctrine. Out of the conflict between the rising interest and breakthroughs in biology, and ever-present dogma, a field of *natural theology* was burgeoning. Had God been using natural (rather than supernatural) forces to shape the development of life itself? This was the question that Darwin become fascinated with, and it allowed him to question evolution without irking the establishment.

In the same way that Broca was not the first to hypothesize about a language center in the left hemisphere of the brain, Darwin was not the first to make suggestions about evolution or natural selection. His brilliance was in synthesizing the patterns of nature that he observed, working with species and their environments as if they were a puzzle whose pieces had been thrown onto the Earth's surface long before.

Darwin, the young and determined naturalist, soon accepted a unique (and risky) opportunity to travel with a ship into the seas. As they charted the geography, he worked as a naturalist, collecting artifacts and making records of whatever he found. As a sign of the times that gave rise to physiognomy, Darwin's epic voyage on the *HMS Beagle* was nearly turned down by the captain because he thought that Darwin's nose was a sign of weak character. The HMS Beagle sailed nearly five hundred miles off the coast of Ecuador to the Galapagos Islands, an archipelago that would be all the laboratory Darwin needed to develop his theory: species evolve over time, and evolution's go-to tool is natural selection.

Darwin's Finches, as they are now known, were the endemic species that he used to showcase his ideas. His theory began when he imagined that all the birds had a common ancestor, and by the process of natural selection, they had become individually fit to their environments (a.k.a. "survival of the fittest"). The beaks of the finches varied in size and shape. Some were high and rounded like a parrot's, some were long and narrow, and some were stout and diamond-shaped. Darwin found that if three types of finches occupied a similar part of the island, one would have a large beak, one medium sized, and one small, which meant that there was only inter-species competition for the various sized nuts and berries on which they fed. Some beaks were better for eating fruit, some for probing flowers, and some for crushing seeds, which led to a mutually exclusive mealtime for them all. The birds that lived in the high tree canopies subsisted off different diets and survival techniques than those that lived on the ground. The birds that didn't have competitors grew because they had free rein over their feed.

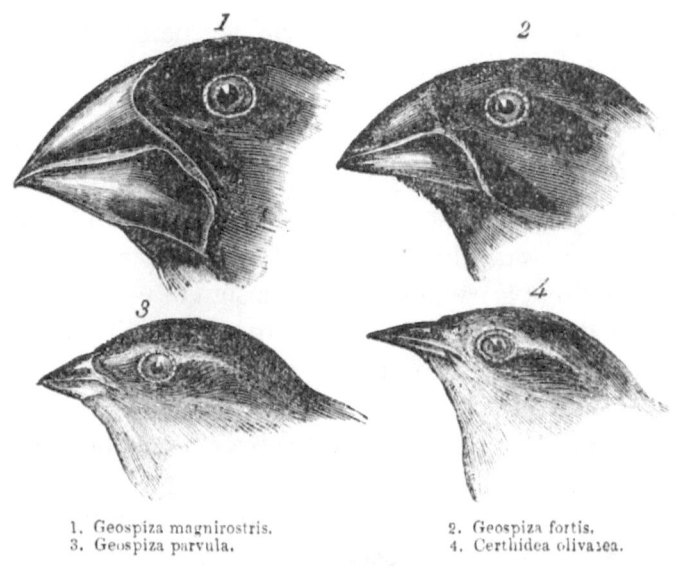

1. Geospiza magnirostris. 2. Geospiza fortis.
3. Geospiza parvula. 4. Certhidea olivaxea.

Darwin's Finches

Biologist Jerry Coyne clarifies the Darwinian mantra by saying, "Natural selection does not yield perfection— only improvements over what came before. It produces the fitter, not the fittest." As the birds became fitter, they became distinctly different, and could no longer mate amongst the whole of the population. Thus, speciation occurred, providing Darwin with a fascinating topic which he would transcribe into his seminal treatise.

Darwin's theory of "descent through modification" suggested that we are not that special in the long history of flora and fauna that have graced planet Earth. In fact,

Copernicus had proven only a couple of centuries prior that Earth was not as significant in the overall scope of the universe as we had thought. The animals over which we were given dominion in the biblical book of Genesis turn out to be our distant kin; you could say, in fact, that some of them had dominion over us until we began our unique divergence only a couple of million years ago.

With this in mind, it becomes rather difficult to accurately define who we really are, and how we are different from other life forms. We are certainly too far removed in our biology to mate with an orangutan and produce offspring, but we must somehow reconcile all of our similarities that have presumably come from a common ancestor. More importantly, how and why was it that our brains have evolved to interact in a world of culture, community, prosperity, and charity? How do we sharply define human capacities and cruelties if we are indeed just another species among a widespread continuum of living organisms?

In 49 B.C., Julius Caesar approached a shallow and unassuming river situated between modern-day Venice and Rome, running from the high western Apennine Mountains that form the midline of Italy, eastward into the Adriatic Sea. Crossing it would solidify the hubris which he had forever lived by, permanently branding him an animal of both war and politics, in search of order among chaos. There was no choice, though, for the pride of a great conqueror, the democracy of a great nation, this river, The Rubicon, had to be crossed. Caesar looked at the legion of men behind him and proclaimed, "Let the die be cast!" Evolution itself is an assemblage of Rubicons, and where we are now is a forward-only progression of where we have come from, with no turning back. Where is our Rubicon? The difference between us and a cuckoo or mandrill lies somewhere along a crook, bend, elbow, or crutch in the great river of life. Knowing where we came from is a forerunner to knowing who we are: the great species that reads books about brains as a pastime.

*　*　*

Finding records of life from one hundred thousand, or perhaps even millions of years ago, is a challenge enough for archeologists, without taking into consideration how one might go about extracting the mind from the bones. The process of fossilization requires that something be swallowed by the sea, or by volcanic ash, which I presume was as undesirable for animals then as it is now, and avoided at all costs. This leads to a fossil record that is certainly missing the majority of life that has existed on Earth, and we may be too arrogant to assume that we know even a minuscule percent of what has come and gone over the past three-and-a-half billion years.

The most common method of dating the remains of life is done by measuring special types of elements called *isotopes* against each other. One of these isotopes decays at a known rate called the "half-life," and the other doesn't decay at all. When compared to each other, their ratio gives a good indication of how long ago the organism died. This method is, however, limited to dating organic matter only to about 60,000 years in the past, because the changing carbon isotope decays so quickly. However, when other inorganic isotopes like uranium are considered from the surrounding rocks or strata layer, dating timescales soar from thousands of years to millions. All dating techniques face deserved scrutiny because they have somewhat complex "calibrations" on which they rely; there is much guesswork, many assumptions, and a healthy dose of relativism (*this* is older than *that*). To drive progress, though, we have to assume that natural forces that form the rings inside of trees, bring in the ocean tides, shower cosmic radiation through our ozone, cause geological shifts, and ultimately bury our ancestors, are not unlike what we experience today. Research is progressing, too, in the direction of putting a carbon-like clock on the rate of mutation within our DNA, thus painting an increasingly clear picture of who our ancestors are and when they roamed Earth, further adding to the sources which can date the mosaic of our past. For now, and specifically in the realm of mammals, we are limited to guessing the overall size of an animal from its longitudinal bones, its stature and mode of movement from the pelvis and spinal cord, its dexterity from the upper extremities, its diet from dental casts, and its mental capability from the cranium.

The search for our story, the pilgrimage from Pangea to the Parthenon, is introduced with great felicity by Aldous Huxley:

> *Surely the mind is the key to our success as*
> *a global species. It is not to be sought in our rather*
> *tiresome torsos nor in our exasperating extremities,*
> *despite the marvels of poise and skilled movements*
> *of which karate-ka, ballet dancers, acrobats and a*
> *few others are capable. Rather it is to be found in our*
> *breath-taking brains and our miraculous minds.*
> *Ladies and gentlemen, may I propose a toast—*
> *to the mind of man. Long may we celebrate how*
> *we cerebrate.*
> ALDOUS HUXLEY

The good news is that we know where to look for the mind; the problem is that the mind dithers from the body as does life itself. This is not to mention the catastrophes, shearing pressures, and chaos that ensue in the process of fossilization. There are, however, two traits relating to the brain that do persist, and they remain integral in any theory of the mind: size and structure.

The endocranial volume (the area inside one's head) has been used by Linnaeus for classifying of species, by Broca in investigating ones intelligence, and now forms the base for nearly all paleoanthropological debates. It seems that with every new generation of science and scientist, we scream blasphemy to the interpretation and context of previous cranial measurements (concerning our kin, race, sex, etc.), and then replace those theories with new ones that may one day be just as outlandish. The path to progress is indeed arduous.

Has any modern correlation been made between volume and intelligence, aptitude, or general behavioral traits? Not concerning *Homo sapiens*, at least. Some research even shows that height may be a better factor of correlation with intelligence than brain size. If bigger were better, then nearly every species of whale, elephant, walrus, and dolphin would be our intellectual rivals or superiors, and the lab mice that run through complex labyrinths would have an all but functionless and vegetative existence. We must also consider the difference between *how smart we are* and *how we are smart*.

Brain size is a reasonably objective way to distinguish between species, but the structure and form of the brain hold the key to what makes us so different (or special); this is the true mark of our evolutionary Rubicon. The story of our river crossing was uncovered in the Olduvai Gorge of eastern Africa, beginning in 1959 by a pair of fossil-hunters.

They were uncovering remains of a hominid whose cranial capacity seemed to fit in between two different species that lived around three million years ago. It was larger than the *Australopithecus*, but not quite as large as anything in the *Homo* genus.

Philip Tobias was the paleoanthropologist who spearheaded the launch of the new species in light of the discoveries, proposing that primate be named *Homo habilis*. His assertions quickly came under fire—some from his cranial measurements, some from his tooth measurements, and some from the sacrosanct status of the *Homo* genus itself. The responders called *Homo habilis* an "empty taxon," said it "should be formally sunk," hoping the species would disappear as rapidly as it came. Out of this controversy came suggestions of five additional nomina with which to classify the new fossils: *Australopithecus africanus habilis*, *Australopithecus habilis*, *Homo Palaeonthropus Habilinesis*, *Homo erectus*, and *Homo erectus habilis*.

At the core of the debate was this: if this species were to be applied to the genus *Homo*, it meant that we (you and I), the *Homo sapiens*, were direct descendants. A recurring theme seems to be that when we become involved—whether the subject matter be a new species, brain localization, evolution, or the discovery of DNA—the stakes are raised, and tension mounts.

* * *

Consider the human brain, defined by its curly neocortex, the seat of intellectual prowess smooshed into and against the dorsal interior of the skull like a wet dish sponge into a narrow shot glass. If our heads were to grow any bigger, we may come to the tipping point of our evolutionary advantage. We would be prone to broken necks and forced out of the moderately-sized birth canal even more prematurely than we already are. Packing neurons into our brain is exactly like stuffing a backpack with too many clothes—eventually you end up with a big, wrinkled mess. Even the sutures that hold together our cranium look like tightly fastened zippers! The protrusions of brain matter are called the *gyri*, and the crevices the *sulci*, while large structures separated by convolutions and fissures form a map of lobes and lobules. From the embryo through postnatal stages, these structures, although soft, make slight impressions on the interior of the still-forming skull. With each beat of the heart, the growing brain's unique topology is imprinted on the brain case. Paleoneurologists call these *impressions gyrorum* and use them as an investigative tool.

The facts concerning language localization that emerged from the eighteen hundreds were useful in investigations of million-year-old fossils. These prepubescent impressions could help identify structures of the brain that are unique to language-enabled animals—structures like the one Paul Broca famously identified, known as *Broca's Area*. Language has long been held as one of the few characteristics that makes us so-human, and it would undoubtedly thrust the bipedal primate from two million years ago into our direct ancestral lineage if it were a speaking creature. Sounds, words, and writing gives us a sense of identity among other life forms through the wonderment of Socratic rhetoric, the playwrights of Shakespeare, and prose of Voltaire.

Now, instead of the purely volumetric interpretation of brain structure that enamored phrenologists and craniologists, the paleoneurologists who gathered the skulls from the Olduvai Gorge were looking for deviations of individual gyri and asymmetries between the brain hemispheres. With specimens from the dig site in hand, Tobias found that the brain volume of Broca's Area in these skulls far exceeded the apparent degree of development seen in the *Australopithecines*. Also included in his examination was another important language center that sits posterior to Broca's Area, named after the neurologist Carl Wernicke, which made it the first time in the history of the early hominids that the two most important neural bases for language abilities appear in the paleoneurological record.

While the case for a unique morphology strengthened with subsequent fossil discoveries throughout the 1970s, reinforcements were needed in the debate. Since language is an un-fossilizable entity, it is actually more proper to talk about the capacity for it, rather than the presence of the spoken word itself. The good news was that support for an undeniably advanced species was abundantly scattered throughout the African ravine. Intermingled with the cranial fossils were hoards of palm-sized lithic flakes. These rocks had been seemingly shaped by unnatural forces and used as tools, hence the final designation of the species, *Homo habilis*, Latin for "handy-man." He was a tool-user, tool-modifier, and perhaps even an ad hoc tool-maker. Even upon comparison with our cousins, the chimpanzees, who are known to use stones to break open seeds and twigs to fetch bugs from small holes in trees, the tooling here was fundamentally different. At the excavation site, there were signs that the materials had undergone a strict selection process. The tools were fashioned rather than foraged, and not only were they creatively employed, but they were consciously harbored for the future. We have yet to come across an animal in our time with this ability and forethought, making the case for *Homo habilis* increasingly reaffirming.

Imagine a colony of upright-walking men, women, and children. The quarries were their tool sheds, home to the first stone masons. Crackles and screams echoed across the battered terrain of the Great Rift Valley as rock was forged into tool, and tool was thrust into prey. Cries of newborns were hushed by a mother's embrace. As the division of labor emerged, the seeds of gossip and camaraderie were sewn. Albeit muffled and subdued, voluntary vibrations of the larynx filled the mild African air. Embedded within them were intonation, context, and a luscious awareness of space and time. Everything from flower buds to flickers in the night sky became all too curious, as ignorance curtailed and culture awakened.

*   *   *

Here was quite possibly the holy trinity of the humanity, the *sine qua non* of culture: bipedalism, tools, and language. This was also the dawn of noticeable asymmetries in the brain; although even Tobias was quick to frame his claims of Broca's Area as only speculative for language itself. He rather used the term, *neural substrate*, submitting that parts of the brain have merely evolved in the directions that have, by chance, introduced evolutionary advantages. Even so, evolution is just a succession of random genetic mutation—it has no agenda, no plan, no direction—and it would be naive to say that language and tooling were meant to be part of our lives.

Consider the analogy of a computer placed in an office for the sole purpose of keeping track of employee time cards. Although its job today is to serve a simple purpose, one day it may be used to manage pay stubs, then organize client information, act as an email server, and eventually evolve into something much different. The transistors etched within the computer's silicon are the homologue to the neurons in our brains, with evolution constantly writing new software, swapping in new hardware when needed, and allowing what is unused to shrivel into oblivion. Sometimes we develop the capacity and skills for something before we know what that something is, we are bearers of *evolutionary preadaptation.*

Psychologist Michael Corballis gives some substance to this theory through three traits of the neural substrate that were developing in Broca's Area: embeddedness, recursion, and generativity. Consider a lone habilis-man peering at a round rock on the ground. He might think there is some potential in that rock; through force, it is malleable, and its purpose varied. He knows that by hitting that rock on another, over and over, it will flake and soon become sharpened. With a honed rock, he could whittle sticks for hunting, cut animal hide for clothing—the possibilities are endless! These realizations are Corballis's traits, and they are seen in language, too. Speaking is an ordered, yet recursive, concoction of phonemes. Sounds lie within words, words within sentences, and while these sounds may be finite based on our physiology and memory, their sequences and pairings are only limited by our near-infinite imagination. Through this, meaning and understanding emerge, thereby defining communication between one life form and another.

It was presumably the advancement of one of these skills that promoted the advancement of the others, and with an ensuing symbiotic explosion. Man did not gain his abilities *à la carte*, they came together. The first atom had been split, the Rubicon crossed, and the evidence humbly recorded in the left hemisphere of fossils from eons ago. As Raymond Dart said, we were coming to the end of nearly "1,000,000 years of gesture and babble," into an era of clarity and comprehension.

The icing had been spread on the cake for Tobias, and it comes as no surprise that *Homo habilis* is now named among the other members of our genus in any textbook or treatise. Henceforth, Man was no longer a slave to the dictates of the environment; it was, rather, the other way around. This was a true transcendence of biology. The tools of *Homo habilis* were the technology that would build great civilizations, and their primitive language the medium through which economies and ideas would be transmitted. The rapture of a haiku, the majesty of the waltz, the romance of a sonata, and the whisper of the divine eddied in the not-too-distant dust.

CHAPTER THREE

Asymmetry

The absence of symmetry seems to be evidence
of the progress of evolution.
— GOETHE

The idea of randomness in biology is just a
reflex.
— JOHN HUBBARD

Thus, the future of the universe is not
completely determined by the laws of science, and its
present state, as Laplace thought. God still has a few
tricks up his sleeve.
— STEPHEN HAWKING

Although Gall, Combe, and Broca had all made particular contributions to the identification of a criminal through physical traits—largely craniocentric suggestions—it wasn't until 1876 that a true theory of criminology based on a man's exterior had been born. This was accomplished by the work of Italian physician and psychiatrist, Cesare Lombroso, who in part agreed with the constructs laid out by phrenology, believing that what is seen on the outside of man is a reflection of what is within. Starting as a relatively small volume of work and later growing into five editions, Lombroso's *Criminal Man* delivered the essential elements of what he would call "criminal anthropology," which took physiognomy into an entirely new domain. He believed that nearly all criminal traits were simply expressions of the savage animals from which we came. The closer one was born to the form and mind of an evolutionary ancestor, the more inferior he was, thus giving him a greater chance of becoming a criminal. This echoes the remark made by the German biologist Ernst Haeckel that, "ontogeny recapitulates phylogeny"—or that the biological development of any organism replays the evolution that it took to get there. It remains an interesting theory, being that we grow as a fetus with a vestigial tail, we shed a layer of thick hair in the womb (called *lanugo*), and that we are developmentally similar to our primate cousins for the first years of our lives. Just remember that the next time you ask a group of children to stop "monkeying around."

The exploding field of biology and the developments with regard to evolutionary theory in the nineteenth century provided the toolsets that Lombroso could apply to the policy and punishment of society's degenerates. The field of criminology had so far been, in the eyes of Lombroso, a mixed bag of archaic and misguided philosophy, with very little application to what he saw as the true afflictions of the people. He had a similar response to interjecting religious dogma into law as Laplace did with regard to his mathematics. When Napoleon asked Laplace where God fit into his mathematical formulations, with the reply, "Sir, I have no need of that hypothesis." And neither did Lombroso.

Lombroso once wrote, "Anthropology needs numbers, not isolated descriptions, especially for use in forensic medicine." The caveat to this statement is that it appears directly after citing jug ears and swollen temples as definite signs of human malfunction. Although Lombroso was dedicated to perfecting his own science, he was enormously tainted by prevailing stereotypes, and was perhaps too quick in accepting the new Darwinism as the force that controlled us all. He was on the very cusp of what would come to be known as *biological determinism*: the inescapable dictates encoded in our DNA.

Following a cranial analysis from 832 living criminals, Lombroso made a variety of absurd conclusions, including the tendency of forgers to have large heads, claims that pickpockets are shorter and weaker than murderers, and that hunched-backs are common among rapists. His biases were instantly extended to anyone who was not a white man, as these individuals were, without a doubt, bearers of smaller (and lighter) brains, and without proper moral equipment. This implicated even children in his obscure theory based on atavism, assuming that they were all corrupted schemers out of the womb. It was Lombroso who popularized the term "born criminal."

Despite having rationale for most of his claims, his one absolute contradiction was in the paradox that if women were, in fact, inferior to men, they should be more prone to criminal acts. However, the numbers didn't add up, and they haven't since. To this day men make up about 90% of the homicides in the United States, and have been the nearly sole instigators in wars and genocide.

In light of racist, sexist, and ageist remarks, Lombroso remained an influential figure both in his time and now. A prolific writer of over thirty books and one thousand publications, he influenced and highlighted the growing need for a more rational and inductive approach to moral policy in France, Germany, England, the United States, Latin America, and even Asia. He was a dedicated humanitarian and perfectly understood how important it was to treat health and poverty in order to reduce crime. He was one of the first people in a public forum to suggest that divorce among a married couple should be made legal based on the evidence that unhappy marriages lead to malevolent behavior—a finding that is still not taken into consideration in some cultures. Although he was without the language of modern geneticists, he was one of the first criminal theorists who set out to prove the phenomenon of epigenetics, and the belief that criminal tendencies and pathologies can be passed from one generation to another.

Many of us would like to condemn Lombroso, and nearly anyone who thinks they can "judge a book by its cover," but this reflection would be best made walking down a dark alley alone at night, rather than cozied up behind a book or computer screen. In fact, it is hard to imagine anyone who doesn't have a slight instinctive reflex to someone who *looks* criminal, and without explanation, we all have our own idea of what that means. To be without this instinct is utopian and arguably too idealistic for the world we live in—it is a survival mechanism, and it has remained in our genes for probably one reason: it works.

Aside from Lombroso's conclusions about a person's brow, forehead height, cranial volume, and nose formation, he notes one trait that is strikingly relevant to the study of laterality: the observation that criminal skulls present frequent asymmetry. In the third edition of *Criminal Man*, Lombroso makes clear reference to the asymmetry found among born criminals, and even offers an album of criminal photographs as proof of the striking abnormalities.

Does this have any relevance to the archetypical image of a thug or convict that comes into your mind? Perhaps they have a slouchy eye, hanging lip, or crooked nose. Although we can agree that most asymmetries reflect malformation, some are celebrated as legend, like Elvis's left-sided lip curl, or model Cindy Crawford's left-sided mole.

Maybe slight asymmetries on good-looking people are just the hint of danger to which we become attracted! Without considering the anomalies, Lombroso seemed to convince himself, and many others, that asymmetry was an utterly despicable trait to lay claim to. But what would Lombroso have thought if he knew about the natural asymmetry in our brains? The evolution of asymmetry that Philip Tobias revealed in *Homo habilis* raised that particular species to higher status, not lower. We assume that the lateralization of our brains is, in fact, a special trait of our superiority; however, it so often signals deformation and error in creation. So what is better, asymmetry or symmetry?

*　　*　　*

In 1928, Paul Dirac was trying to mathematically marry quantum theory and Einstein's theory of relativity. He was successful in explaining one of the atom's "last puzzles," coupling the dynamics of the fast-moving electron as it circled the atom's nucleus in a way consistent with relativistic theory. Dirac's solution, however, was anti-communicative, meaning that it revealed two opposing solutions for the particle in question, much like the way in which the quadratic formula we all learned in grade school always ends up with both a positive and a negative result. In the case of Dirac's equation, the first result explained the existence of the well-known, negatively-charged electron. At first, the second solution appeared to be a mathematical artifact; there was no known particle that was identical to the electron, yet opposite its charge. For this inexplicable bit of his mathematics, Dirac was subject to brutish criticism during his lectures, at which he had no other way to explain this imaginary particle than to call it an "anti-electron." But it wasn't long before the two-solution equation got its

Picasso-status: in 1932, the very antiparticle he predicted was found in a cosmic shower. Dirac's paper-and-pencil work would flourish and be applied across the world in high-energy particle labs throughout the next half-century, when the existence of antiparticles would be thoroughly proven. To the great quantum physicist Werner Heisenberg, this discovery was "perhaps the biggest jump of all the big jumps in physics in our century."

The symmetry of Dirac's particles and antiparticles is so perfect that if they ever meet, they instantly annihilate each other—poof! It is this romantic suicide pact that preserves the precious law of conservation that says for every "+1" in the universe, there is also a "-1." Once upon a time these conflicted particles were contained within the singularity that gave rise to the Big Bang, and entity of equilibrium, on the first day of our cosmic calendar. Science still does not answer why there was a singularity, why it exploded, and why matter was allowed to turn in these opposite directions, giving rise to our worlds of matter, and the hypothetical distant worlds of anti-matter. If this was God's intervening finger, letting there be light, we should be reminded that he did so with great symmetry.

Since it appears that symmetry was decided in the beginning, it should be no surprise that symmetry itself has historically been a synonym for, and symbol of, the gods. This balance has been held in such high regard that asymmetrical elements have purposefully been built into ancient art and architecture in order to not encroach on the perfection of the gods. This has been a dictate of Islam—to always include a deliberate error into artwork in respect of the perfection of Allah—which you can test the next time you set foot on a Persian carpet. Sometimes these asymmetries are built into twin towers of churches, deeming the larger of the two "Adam" and the smaller "Eve." Although this can be purposeful, it sometimes does truly reflect the doing of an imperfect species. Most famously, the towers of Notre-Dame in Paris are different heights, solely due to the project being swung carelessly from one architect to the next.

In one of his lectures on asymmetry, Richard Feynman tells the story of a famous bridge in Japan, relating it to the natural predisposition of placing asymmetry into the world around us:

> *The only thing we might suggest is something like this: There is a gate in Japan, a gate in Nikko, which is sometimes called by the Japanese the most beautiful gate in all Japan; it was built in a time when there was great influence from Chinese art. The gate is very elaborate, with lots of gables and beautiful carvings and lots of columns and dragon heads and princes carved into the pillars, and so on. But when one looks closely he sees that in the elaborate and complex design along one of the pillars, one of the small design elements is carved upside down; otherwise the thing is completely symmetrical. If one asks why this is, the story is that it was carved upside down so that the gods will not be jealous of the perfection of man. So they purposely put the error in there, so that the gods would not be jealous and get angry with human beings.*

> *We might like to turn the idea around and think that the true explanation of the near symmetry of nature is this: that God made the laws only nearly symmetrical so that we should not be jealous of His perfection!*
> RICHARD FEYNMAN

49

Feynman prefaces this anecdote by asking, "Why is nature so nearly symmetrical?" In apparent agreement, G. K. Chesterton, who stood outside the world of quantum wonder, remarked that perhaps the world "looks just a little more mathematical and regular than it is." And of course, J. B. S. Haldane's quote will forever resonate on this note:

> *My own suspicion is that the Universe is not*
> *only queerer than we suppose, but queerer than we*
> *can suppose.*
> J. B. S. HALDANE

In 1956, a famous experiment at the quantum level proved that we live in a world which, in fact, knows right from left, and if everything were to be mirrored, so-called "parity" is not conserved, making a reflected world not equal. What this seems to suggest, whether you choose to follow theology, physics, or both, is that there is a deserved reverence for symmetry (and asymmetry!) in the world around us. However, there are small disputes that need explanation, things that are positioned so far from cosmic calendar day number one that we must consider if God really did start with perfect symmetry when matter was split, and if we truly have become part of a fallen world since.

* * *

Whether or not it is a falling from perfection, asymmetry is an essential element in the building blocks of life, amino acids. These are what build and regulate our cells and tissues, and it is clear that if an asymmetry did not exist, we wouldn't, either. In 1849, Louis Pasteur discovered that two types of tartaric acid—one organic and one artificial—were nearly identical in chemical structure but had very different properties. The compounds were constituted of the same components; it was simply that the carbon bonds were reversed down the midline (said to be *chiral*). Depending on which side of the midline the carbon appears in the compound, it is named with a preceding "D" or "L" (taken from the Latin roots of *dextro* for right, and *laevo* for left). Any trip down the supplement aisle of a health store will give you an idea of the handedness that our bodies rely on: L-Arginine, L-Carnitine, L-Tyrosine, L-Taurine. Every living organism is just like us, relying on left-handed amino acids and right-handed sugars. This is why those supplements all begin with an "L," and why naturally-occurring sugar is called "dextrose." It turns out that the shape of such molecules plays a game of lock-and-key with our bodies, making only some of them useful and digestible. For instance, although L-sugars taste similar to D-sugars, they are not absorbed into our bodies. Tricks like this have spawned the booming market for zero-calorie artificial sweeteners.

Finding organisms composed of D-amino acids is kind of like finding antimatter—it's so scarce that it's barely worth mentioning. Even meteorites from deep space show that molecules from distant worlds share a similar handedness to earthly life forms, suggesting that our particular paradigm is consistent with, at the very least, our solar system and likely, our entire galaxy. When we send astronauts or robots into outer space to collect samples of rock from the moon or Mars, this is the type of thing for which scientists test.

There are several theories of how and why asymmetry was ever introduced into the natural world in the first place. Some point to polarized light, others to nuclear forces; however, any consensus on the origins of asymmetry is far from overwhelming. It remains a cosmic mystery that is embedded deep within us all.

One thing that we can say is that our world is probably not "special." That the molecules have taken one side, or that electromagnetic forces spin one way, appears to simply be a choice that had to be made at one point in time, probably within the first seconds of our cosmic calendar. It is the old problem of Buridan's Ass. Jean Buridan, a French priest and celebrated figure of medieval science, outlined a predicament in the margins of some Aristotle that he was reading. If an ass is placed in the middle of two piles of hay, how will it choose which one to eat? We can't blame the ass for breaking the symmetry of the situation—otherwise it would starve—but how can we explain the choice it made between the two identical piles of hay lying equidistant from its position?

Apparently, Mr. Einstein, God does play dice! Many psychologists, including Freud and Skinner, did not believe in free will and would argue that all we can do is respond to internalized or externalized stimuli. Until we have a complete understanding of how a deterministic system would operate, down to each subatomic particle, we can only suppose that we have the opportunity to choose our own actions. Whether or not free will is granted to Buridan's Ass, we know that living another day is important to any ass, and it will eventually side with the left or the right by some means or another. The cosmos must always be making these decisions, which could be the greatest proof for an underlying asymmetry—albeit chaotic. If the world never "gave way" to one side or the other, how would galaxies know which way to spiral?

* * *

The greatest and most resonating insight into the question of whether asymmetry is better or worse comes from the study of our phylogenetic relatives, and from ourselves, as we exist in nature. So what does asymmetry actually boil down to in ecological terms? For an individual, it is simply the value of the right feature minus the left. You can test this yourself by measuring your right arm and comparing it with a measurement of your left arm. However, in a population census, asymmetry is instead measured by the variance of traits in a given geographical boundary; that is, how much your arm measurement varies from those of your friends.

These measurements turn out to be important indicators of the health of a person or group of people. From the moment of fertilization through adulthood, all organisms are subject to stresses that can shake the natural state of homeostasis—the comfort zone—and reduce the developmental stability that guides the growth plotted by our genes. Our trajectory can be skewed by the pollution around us, the food we eat, the temperature of our environment, and even noises if they are loud enough. These are the perturbations that we either cope with, or succumb to, but that which we are ultimately shaped by.

One of the easiest (and cheapest) ways to monitor the health of an environment or habitat quality is to measure the levels of asymmetry found in its constituents. The appearance of, or rise in, asymmetry is a sign of trouble. For instance, asymmetry increases in the leaves of trees that surround copper and nickel smelters which emit aerial pollution, and the distance a farmer's crop is planted from large power lines has been shown to be proportional to the level of asymmetry found in the end product. Flowers with asymmetrical shaping are less attractive to pollinating insects, resulting in fewer seedlings. This is a phenomenon to which not only plants and crops are subject. Children with alcoholic mothers who suffer from fetal alcohol syndrome have increased physiological asymmetries, as well. As a general rule, there is a negative correlation between asymmetry and fitness throughout the entire kingdom of plants and animals.

This trend toward an asymmetrical disadvantage can be viewed teleologically, as well, putting aside the nature of the cause altogether. For many reasons, it is well established that symmetry—or the lack thereof—plays an important role in survival. It offers an obvious "mechanical" advantage that keeps our bodies in equilibrium when we have to react to danger. One can imagine trying to run or punch if one leg or arm were much larger than the other. Of course, the situation is more dire for animals like birds and fish, which would be completely incapacitated by a major asymmetry in their wings or fins. Several studies involving the common house fly and the dung fly show that the risk of predation rises in direct relation to their asymmetrical features. In full corroboration, a staged lab fight between two male damselflies found that the fly with more asymmetry had a consistent disadvantage.

There is a corollary that I think is worth mentioning, which brings to mind a Wave Runner I rode in the lake as a child. There is a coiled plastic safety lanyard that you are supposed to have attached to your body, which shuts off the engine if you fall into the water. In case you forget to wear that, there is a spring mechanism that pulls the steering column slightly to one side at all times. Instead of losing control and watching the machine idle off into the horizon, it will do moderately sized circles in the water, allowing you to catch it at some point. This strategy seems like it would be advantageous for us, as well, if we were ever found in a "default mode" or unconscious state. We would never want to venture far to find help—a straight trajectory may lead us farther from our base camp, stop others from finding us, or foray us into unknown predatory zones.

Lo and behold, something similar actually happens in the animal kingdom. Locusts (and several other flying insects) have been found to have an asymmetry in the motor nerve output that connects to their flight muscles. During normal flight, their visual feedback causes a single-sided double-firing in the neurons of their central nervous system, which results in flying straight. However, when this system is damaged, single-firing is reinstated and results in a spiraled path.

This phenomenon realizes itself in us humans, as well. Ernst Mach, who gave us the famous scale by which to measure shock waves, learned from a retired army officer that on dark nights or in snowstorms troops will move approximately in a circle. In scientific studies on the matter, subjects have been blindfolded (or placed in fog) and then asked to walk, swim, or drive in a straight line. The result is that they consistently end up going in circles. We can only make assumptions about whether this has been a carefully selected trait, or is just another sign of developmental instability, but the asymmetry is curious indeed.

Not only is the symmetry of our body parts important for survival, but the mere fact that they are redundant is important, too. Apparently, two really is better than one (and three is far too many). As in the case of the bird's wing, fish's fin, and horse's leg, an animal's survival counts almost entirely on that functional, bilateral partnership. It is also important to consider these counterparts as weaponry, and to not forget how disadvantaged a deer would be without both of its antlers, a walrus without both tusks, or even ourselves if we were unable to fight with two hands. However, most animals can lose limbs, appendages, or senses as long as this doesn't happen to both the right and left sides. This theory of duplicates breaks down, to some extent, for the internal organs, but it could be reasoned that the complexity of having two hearts and their visceral accoutrements outweighs our needs; most bodily punctures become life-threatening by means of internal bleeding or infection, to which the redundancy would offer no relief.

The crossbill is a special bird from the finch family that happens to be asymmetric on an individual level, yet symmetric on a population level. This situation of symmetry is known as *antisymmetry*. Crossbills are appropriately named for their beaks, which overlap and fold to opposite sides when closed, making them specialists at extracting seeds from pine cones. As their beaks come together, they create a twisting action—like moving your fingers past each other when twisting a bottle cap—thus displacing the seed and plopping it directly into their mouths. Their efficiency at this task has carved them a unique ecological niche. Most interestingly, and the reason why their population variance flatlines, is that all European crossbills have their upper mandibles pointing toward the right, whereas the North American variety have it pointing to the left. It seems that there was at one time a common ancestor, but that the specializations formed distinctly in each landscape. The pines that provide the crossbills' nourishment have an asymmetry, as well, found in the grain lines that run up and down their trunks. While it would be an interesting causation for these trees to be separated by the same geography as the crossbill's beak, their spirality is most distinctly separated by the equator.

Another bird substitutes antisymmetry for a strict asymmetry. The wry-billed plover's beak turns slightly rightwards, helping it turn over stones and look for food. Why are there no left-turning beaks? It could very well be a story similar to that of our own. Birds learn almost entirely through mimicry, which would mean that the side to which Mama Bird chose to turn her head would likely be the side that her fledglings chose, also. Likewise, there exists more than one way to tie your shoes, but you probably do it the way your parents taught you. The second possibility is that an asymmetry in the wry-bill's muscular structure developed in favor of the right-beaked birds—a theory that we humans also have to consider for ourselves.

Of the other few antisymmetries found in nature, there is that of the male Narwhal whale, whose left—and never its right—tooth protrudes nearly eight feet out of its upper jaw, making it the closest thing in nature to a unicorn.

The normal distribution for left and right traits takes the shape of a single-humped bell curve, representing a population that is mostly the same. There can be *directional selection* that shifts the entire population's traits; and in the case of the crossbill, there is *disruptive selection*, which results in a bimodal (two-humped) curve, representing a split between two equally effective methods of survival. The marine skate (very similar to a manta ray) has such a strategy to evade predators, choosing to either lay low buried in the sand and let a predator pass it by, or to rapidly move away at first sight. As you can imagine, anything in between is simply not effective. Beyond behavioral bimodality, this condition is biologically manifested in the *Uca musica*, a small fiddler crab named for its acoustical performances when attracting a mate. Although born with two large "male" claws, one is either damaged or eventually falls off early in life, leaving the crab with a small "female" claw in its place. This occurrence flips a biological switch, and no matter how many times the crab loses and regenerates its claws after that point, they will always grow back the same size (one big, one small).

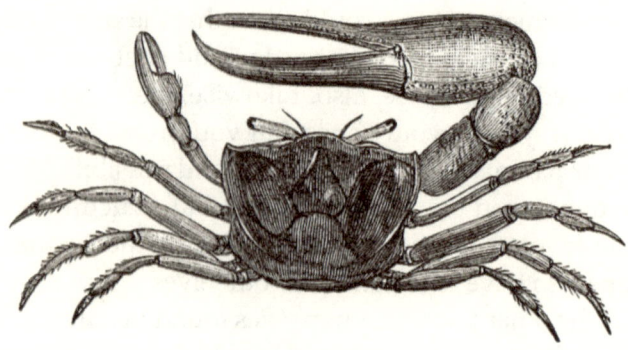

Fiddler Crab

In a similar fashion, the flatfish loses a piece of cartilage between its eyes at an early age, allowing both eyes to drift to one side, with a fifty-fifty chance of them progressing either rightwards or leftwards. They have been able to maintain this as an evolutionary advantage, allowing them to swim horizontally along the sea floor and still see with both eyes. Most female birds are also harborers of this type of unique transformation in that their left ovaries are usually the only functional organs, and the right remain vestiges. If the left ovary is damaged or removed, the right one comes alive and becomes even larger than the left had been.

This split of traits between right and left even affects some animals in their sex lives. Take the Anablep, a famous fish which appears to have four eyes, two that sit above water, and two below. This odd creature is born with its sex organs on either the right or the left side, the distribution of the trait apparently as random as a tossed coin. This means that it is forced to find a mate that is not only of the opposite sex, but that also has its sex organs on the corresponding side!

Symmetry additionally seems to play with our senses and perception. We have machinery inside of us that can quickly identify symmetry allowing us to completely remove focus on the whole of an object, and narrow our focus onto only one side, thereby reducing the amount of information we must process. Our emotional tie to symmetry in art and form may very well be a result of our brains slipping into a zen-like state of mind, finding comfort in reflected objects. In more than just humans, symmetry is a crucial component in "sizing up" a mate and stands as an incredibly universal classifier of beauty. After all, asymmetry is (apparently) biologically deleterious. Ideally we find a mate who is healthy, fertile, and able to both fight and forage for offspring: all things that have been shown to strongly correlate with symmetry.

It appears that, in many ways, Cesare Lombroso was onto something. While asymmetry is a feature that peaks with our species—the most intelligent, diverse, specialized, culturally attuned, and dominating creature ever to grace Earth—nature seems to signal something quite different and glaringly ominous. It should be alarming to us that there are only a few animals that occupy a niche of asymmetry, finding refuge on the fringes of normalcy. If you listen ever so faintly, there is something out there crying, "retreat, retreat..." back to a place of balance and stability, back to our symmetrical origins.

CHAPTER FOUR

Handedness

Our puppet strings are hard to see,
So we perceive ourselves as free,
Convinced that no mere objects could
Behave in terms of bad and good.

Are you, our transcendental gods,
likewise dangled from your rods,
and need, to show spontaneous charm,
some higher god's inserted arm?

We seem to form a nested set,
with each the next one's marionette,
who, if you asked him, would insist,
that he's the last ventriloquist.
— THEODORE MELNECHUK

* * *

The most obvious right-and-left opposition should be evident even as you turn the pages of this book, and surely the next time you pick up a pen. Have you ever wondered why you and most all of your friends are right-handed? This bias contributes to our right hands' being stronger and more learned in both delicate and complex tasks, leaving our left hands immature in many regards. Other mammals also develop hand preferences, making tests for the mannerism common in animal laboratories that perform experiments on one of two hands, feet, or brain hemispheres. By knowing which hand an animal prefers, research teams can make assumptions about which side of the body will be more active or fruitful for investigation. A striking fact is that the right-left preference distribution of these animals is around 50-50, whereas the right-left preference of humans is about 90-10. These numbers hold true anywhere on Earth, even in removed and remote areas.

The Edinburgh Handedness Inventory was created in 1971 at the University of Edinburgh, near the very streets on which this book opened, only 140 years later. It has remained the simplest method of assessing handedness ever since. Assess yourself: what hand do you use for the following tasks?

1. Writing
2. Drawing
3. Throwing
4. Using Scissors
5. Brushing Teeth
6. Using a Knife (without fork)
7. Using a Spoon
8. Sweeping with a Broom (upper hand)
9. Striking a Match
10. Opening a Box (lid)

So why do we continue to be persuaded into using one hand rather than the other? The origins of handedness extend well beyond cuneiform and intelligible record-keeping, and remain a predisposition that has persisted in us with great gravity. If it were of no consequence to us or to our surroundings, the habit would surely be shaken off with each new generation. However, it appears that we have found ourselves occupying a niche upon which no species has stumbled, or been able to explore. But where did it start and why? Could the sword and shield theory be correct? That we have reacted to threats asymmetrically with an astute visceral awareness, protecting our hearts on the left with a shield, and thrusting a sword with the right? Maybe it is ingrained in our physiology, and we come out of the womb with a fixed tendency to grasp with the right hand? Paul Broca asserted that, "We are right-handed because we are left-brained," but there is some evidence of the antithesis: that we are left-brained because we are right-handed.

In the United States, handedness is more or less an unobtrusive phenomenon, but it carries more weight in other cultures around the world. In 1984, the *Smithsonian Magazine* caused a stir with an article entitled "Nomads of the Desert," which spotlighted Saudi Arabia's Bedouin culture and people. Six pages into the story, there is an image captioned, "Shammar welcomed Easteps to An Nafud with kipsa of lamb, rice, cucumbers, raisins and tomatoes." The picture features a group of nine men all reaching with their left hands toward a tray overflowing with food. Because of the taboo in Muslim countries surrounding right and left eating etiquette, this was immediately called out by the Arabic community, making an otherwise harmless photo-edit (flipping the image horizontally) a major insult to Bedouin society. Certainly no individual would think of using his left hand to eat!

A more famous example can be found in the lithographed version of *The Potato Eaters* by Vincent van Gogh. Not accounting for the inversion that takes place as a lithograph goes through printing, his sketch of five right-handers became a print of five left-handers. In a letter to his brother Theo, he hastily points out his own error:

*If I make a picture of the sketch, I shall make
at the same time a new lithograph of it, and in such
a way that the figures, which I am sorry to say, are
now turned the wrong way, come right again.*
VINCENT VAN GOGH

The Potato Eaters by Vincent van Gogh

It could very well be that the right-left bias is a movement toward an order and systemization that would be impossible for an ambidextrous or equally partitioned group. When asked to mobilize and march in order, there would be no naturally leading leg. Soldiers with arrows, spears, swords, or pistols would interfere with each other when in a tight formation, especially if they had to draw their weapons. We have become such experts in building machines and replicating processes that perhaps through osmosis we have mechanized ourselves—*Homo mechosis* seems just as appropriate as *Homo sapiens*.

One place that is sure to be indicted in the controversy is the classroom, where most kids begin learning about good and bad, alongside the concept of right and left. It was once common in certain parts of Japan for left-handed pupils to receive a beating until they started to conform to the ways of a right-hander. One response to the epidemic was Japanese psychologist Soichi Hakozaki's book, *Warnings Against Rightist Culture*. Children have been found in the Dutch East Indies with bound left arms in order to force "proper" use. Even the Boy Scouts organization has adopted a left-handed handshake, said to be an intentional compensatory scheme enlisted by Lord Baden-Powell who was striving for a more two-handed society (a popular movement in the late nineteenth century).

A study conducted in 1970 surveyed students' handedness in over 1,500 classrooms in Vancouver, British Columbia, and resulted in a shocking truth about the differences between public and Catholic schools. The study found that Catholic schools were noticeably more right-handed than public schools, which could easily be correlated with more disciplined instruction. Many schools have furnished their classrooms with desks that can only be entered from the left, and that provide an armrest and tabletop on the right side. You may also have noticed that the contours on most scissors are designed for right-handers, and that lefties have to deal with a spiral or flap impinging their trajectory when writing in a notebook.

It has been found that forcing a left-hander to be right-handed can result in general confusion, dyslexia, and stuttering in up to 70% of subjects, making the coercion of handedness irresponsible and objectively unjust. The leading hypothesis on this issue has been investigated by scanning the brains of stutterers for activity while they perform speech tasks. While typical people only activate one side of their brain when speaking, stutterers tend to activate both sides. It is very possible that forcing handedness somehow interferes with a necessary coordination of the brain's two hemispheres and conflicts with the normal processes behind speech. It has also been shown that left-handers can recover from aphasic strokes (loss of speech) more quickly, perhaps for the very reason that both hemispheres are primed for generating speech, and thus either one can take over if the other is damaged.

There is no doubt that the world at-large is built for right-handers and most everyday tasks are easier for dextrals. Pianos, guitars, and violins are all naturally suited to the right-hander. Screws drive into material in such a way that utilizes the much larger forearm flexors rather than the extensors. Most written languages proceed from left to right, which keeps the palm of the writer off the newly laid ink. The stem of a watch is built to be manipulated with the right hand. Most instruction manuals depict only right-handed assembly. Wedding rings are applied to the left hand so that they do not interfere with shaking hands and chores. Check your closet on this one: the buttons on a man's dress shirt are on the right side, while those on a woman's dress shirt are on the left. This is because women used to dress men; therefore, the buttons were designed for their right-handed ease. How long have we lived in a world that is made easier for the right hand?

B. F. Skinner was one of the first scientists to make use of operant conditioning boxes that trained animals in successive and repetitive trials until a behavior was learned. He showed that given the correct protocol, animals could be trained to do nearly anything. His findings made us humans question whether or not we had free will at all. Skinner also researched the effects of positive versus negative rewards, and found that we are more robustly influenced by positive rewards than negative. This is why we have all heard that it is so important to reward your pet for good behavior, while scolding him or her for bad behavior is of little use. One cannot go without wondering if what we live in is a gigantic operant box, and if since birth we have been trained that using the right hand makes everything easier. Our right hand results in quicker rewards, whereas using the left hand is awkward, and sometimes results in painful accidents.

The behaviorist J. B. Watson said that following birth, "Society soon thereafter steps in and says, 'Thou shalt use thy right hand.'" Another behavioral attribute which is difficult to reconcile is that right-handers are found to be more robustly right-handed than left-handers are left-handed. That is, right-handers usually use their right hands for nearly everything, but left-handers have a tendency to sometimes use both hands, or have very little preference altogether.

A behaviorist's theory of environmental pressures is corroborated by the many efforts put toward trying to breed a particular handedness into mice. Mice will gravitate toward using one preferred limb when trained to do precise movements. When these mice are separated by handedness and then breed, the offspring have no statistically significant signs of bearing their parents' hand preferences. There does exist some evidence for a so-called "right shift theory," which implicates specific genes in biasing the body's asymmetry and overall handedness; however, the gene coding can only be proven to be an indirect association. The entire premise has to be questioned in light of research that has shown no relevant correlations from the handedness of identical twins, who of course bear the exact same DNA.

Blaming our handedness on an ingrained pharmaco-logical asymmetry has even been suggested as a reason for our natural bias. In fact, it has been ensnared in nearly every aspect of human culture and behavior by Fred Previc in his book, *The Dopaminergic Mind in Human Evolution and History*. Previc's exposé, and other literature, suggest that a prominence of the neurotransmitter dopamine in the left hemisphere leads to an asymmetrical bias in turning, speaking, addictive behaviors, and motor programming.

Despite some evidence that genes or chemical imbal-ances may play a part in our asymmetry, the root cause is almost surely epigenetic in nature—it exists outside of our bodies—because otherwise we would expect to see our asymmetrical attributes more often in our primate ances-tors. In the 1950's, primatologists stumbled on this type of cultural diffusion in a secluded group of macaque monkeys off the coast of Japan. The researchers supplied the shores with heaps of sweet potatoes and wheat for the animals to eat. If you have ever had a picnic at a beach, you know that one problem with such a meal can be sand getting into your food, which was precisely the same predicament and discomfort that the macaques found themselves facing. A one-and-a-half-year-old macaque that the researchers called Imo was the first to solve this problem; she took the food and briefly submerged it in a nearby stream, remov-ing the sand altogether. "Washing," as we would call it, was an entirely new innovation for this clan of primates. The technique was soon adopted by Imo's closest playmate, and then by the other youngsters in her family, and finally, the entire island of macaques was using this technique to clean their food. From that point on, newborn macaques were raised in a society that cleans its food. What Imo had done was something incredibly novel, but it soon became commonplace, with one generation following the ways of the generation before it.

In the past one hundred years, some statistics have shown that left-handedness is on the rise, increasing nearly 10% from the year 1900. Curiously, this rise correlates with a rise in the use of typewriters and computers, which are an ambidextrous means of communicating, ridding us of pesky notebooks and removing the obscurities of the left-to-right hand movement. One explanation for this phenomenon is that society has become more accepting of minorities as a whole over the last century. This is a broad statement, yet in nearly every aspect of human affairs, we have given more opportunity and rights to people who were once oppressed. An equal playing field has been in construction for blacks, women, the disabled, and transgender communities now for decades. A freer and more liberal social structure may very well allow lefties to use their natural hand in everyday tasks. We have to wonder if there is a fast-approaching asymptote to the rise in left-handers, or if we truly are on a course toward the equanimity of our hands, joining the habitations of all the other animal groups on our planet.

*   *   *

The era of classical antiquity in Greece is the foundation of much that is modern today. Literary forms were mastered, architecture was refined, and records of human life on Earth were recorded on slabs of stone and in clay with an explosive vibrancy. During the eighth century B.C., Hesiod became, quite possibly, the first-known person to present mythic and religious dualism between man and woman in the story of Pandora, the first female god in the genealogy of Zeus. As the story goes, her curiosity got the best of her when she opened a jar (or "box") that she had been told not to touch, containing all the evils of the world. As a result, she and her offspring became damned and were labeled a great bane to men. Not surprisingly, the story of Pandora has many parallels in later cultural narratives, most famously mirrored by the account of Eve in the Garden of Eden.

Stories like this, though, were slowly losing their hold on the minds of the Greeks as a satisfactory explanation of the natural world. Out of this curiosity, stepping beyond myth, grew the disciplines of basic mathematics and natural philosophy: investigations of the world through numbers. Instead of characters and gods that had thus far been used to make sense of the apparent dualisms that surround human beings, the first hints of rigorous classification based on numerals peaked. The most exciting work came from the traveling enigma, Pythagoras. Well-known for his theorem concerning triangles, he was more prolifically a leader of thought and even of his own religious sect. He regarded numbers as *la clef de voute de l'Univers*, or "the keystone of the universe" and not only studied them, but integrated them into his worship. He had a deep appreciation for the mysteries of the cosmos and worked towards solving them. In the tones of nature was a mathematical harmony, while at the same time, blowing dust and swirling motes harbored *anima*, or inanimate souls that guided the apparent life within the eddies. The people who prescribed to this doctrine—the Pythagoreans—drew upon a table of ten principles, two elementary columns of dipoles. They were to conclude that substances are composed and fashioned out of these underlying elements.

This table of dichotomies, called the *sustoichia*, may have been constructed by Pythagoras, or perhaps his most

prodigious followers; however, the concepts are a cura-
tion of ideals and beliefs that were widespread throughout
Mediterranean culture at the time. They put a logical, yet
underlying, religious subtlety to those things which con-
struct our world: quantity, sequence, unity, direction, sex,
force, motion, contrast, death, and form. If it were not for
Pythagoras, such right-left dogma may never have passed
into the world of natural philosophy, otherwise known as
science.

Limit	Unlimited
Odd	Even
One	Plurality
Right	Left
Male	Female
Rest	Motion
Straight	Curved
Light	Darkness
Good	Evil
Square	Oblong

What were not present in literature before the Pythagoreans were intentional liturgical inclusions and codifi-
cations. Counting the words, lines, columns, syllables, or
verses of a piece of writing was quite common throughout
Greece, a practice called *stoichiometry*. At first, it was not
done in any attempts to express artistic or philosophical
ideals, but for several practical purposes: it ensured that
texts would be copied correctly, it provided a system by
which scribes could charge their customers, and finally, it
was a way for authors to plan ahead when fitting their work
into standard-sized scrolls. Not until the age of Pythagoras
was the arrangement of such counts and measures coinci-
dent with nature. Writing became all too human, as lines

of text were paired with a musical scale, the juxtaposition harmonizing the words into an organic symphony with letters dancing off the page in perfect pirouettes. The patterns and complexities contained in these scores is now used for chronological dating, in assessments of genuineness, and for retracing geographic origins.

Found within the writings of the most celebrated Pythagorean student, Plato, is an incessant reliance on splitting pieces into twelfths. In *Symposium*, Plato narrates a party of many guests who came to speak on Love. Most recognizable to the modern scholar is the Socratic dialogue which professes the love of wisdom—the literal meaning of "philosophy." Jay Kennedy at the University of Manchester has uncovered the most recent evidence of these harmonic scales and the fundamental role they play in the writing:

> *In the Symposium, Pausanias' speech, Eryximachus' speech (including the repartee over Aristophanes' hiccups), and Aristophanes' speech are each about one-twelfth of the dialogue. Socrates' long speech, including his conversations with Agathon and Diotima, occupies three-twelfths or one quarter of the entire dialogue. Alcibiades' speech lasts about two-twelfths of the dialogue.*
>
> J. B. KENNEDY, PLATO'S FORMS, PYTHAGOREAN MATHEMATICS, AND STICHOMETRY

Plato also architects the lines of *Apology, Protagoras, Cratylus, Gorgias, Republic,* and *Laws* into convenient multiples of twelve. The number twelve is useful in music as a scale because it establishes harmony between two notes with similar multiples. It is often found in Plato's writing that passages of positive concepts are found at intervals that are in accordance with harmonic octaves, negative concepts aligned with inharmonic octaves, and neutral concepts placed in-between.

Beauty was ingrained in their arrangement, embedded in their composition, all while hidden from plain view; the words were not attributed to the supernatural, nor were there claims of unworldly inspirations. It was believed that in following this form, the writing was in sync with the cosmos, and from the words emerged a higher order and deeper understanding. After all, why should this honor only be bestowed to a few privileged prophets?

As to why the facts surrounding the numerology of the ancient Greeks is so vague, it is largely because the writers expected those who were part of their circle to know what to look for, and those outside to be blind to the order. Philosophers were playing diviners, and membership to the club was restricted to those who were seeking answers by means of introspection, rationale, mathematics, measurement, and logic. This tradition would unintentionally evolve as an influence on groups the most distinguished groups of thinkers and taint explorations into the physical world, which were supposed to be entirely objective. It affected a cardinal rule of science: to seek truth without a priori assumptions.

Plato's greatest student, Aristotle, kept the best record of Pythagorean influence on philosophy:

> *Contemporaneously with these philosophers and before them, the so-called Pythagoreans, who were the first to take up mathematics, not only advanced this study, but also having been brought up in it, they thought its principles (archas) were the principles of all things.*
> ARISTOTLE, METAPHYSICS

He adds that they insisted on "the whole heaven to be a musical scale and a number," and was very much aware of the dichotomous associations they stood by:

> *Evidently, then, these thinkers also consider*
> *that number is the principle both as matter for things*
> *and as forming both their modifications and their*
> *permanent states, and hold that the elements of*
> *number are the even and the odd, and that of these*
> *the latter is limited, and the former unlimited; and*
> *that the One proceeds from both of these (for it is*
> *both even and odd), and number from the One; and*
> *that the whole heaven, as has been said, is numbers.*
> ARISTOTLE, METAPHYSICS

Although Aristotle was not a Pythagorean in a strict sense, it was inevitable that some of the reigning philosophies of his day serve as a foundation for his philosophical base. Indeed, he called on aspects of dualism in his teachings: rather than numbers which corresponded to good and evil (or right and left), he assumed everything could be explained by matter and form. That is to say, everything we interpret is formed matter. If a page is matter, then this book could be the accompanying form — though many things could be made of pages, and our contemplation of these endless forms is what life (and happiness) is all about. The concept of forms was, of course, championed by his mentor Plato, but Aristotle's contribution was important because it segued into the philosophy of the mind, allowing mind and matter to systematically disassociate within a dualistic framework.

One aspect of the natural world with which Aristotle was very concerned was how and when the decision is made about the sex of a child. The existing medical literature on the subject associated the right side of the body with manhood, and the left side with womanhood. It was said that the right testicle of a man carried the seed to create male children, and the left testicle carried the seed for female children. Added to this, a male fetus grew on the right side of the womb and a female on the left side. Although these ideas were presented by astute figures of antiquity like Hippocrates and Anaxagoras, Aristotle had every reason to dig into these bold claims, especially because of his particular interpretations of dualism.

Aristotle had studied both human and non-human anatomy and physiology, creating vast catalogues that would support his own theories of embryology and evolution. Through his exploration, he found no sense to be made of the right-left proclivities presented before him—not in man, and not in any other animal. He was keen to point out experiments that removed the right or left testicle of an animal without noticeable results, and went to lengths to debunk any claim that this was a reliable determinant of the offspring's sex.

Aristotle may have been firm with his position on biological growth and development, and clung to empirical observations, but he was not without his own beliefs about the natural cosmic order. He devised four basic qualities of matter, which in reality were two pairs of opposites: hot and cold, wet and dry. Although he all but disposed of the apparent male-female polarity during birth, he still held that there was a superiority of the right side, corroborated by his observations of how men carry burdens, how they walk, and the hand with which they fight. He associated the top and the front along with the right, stating that, "Above is more honorable than below, and front than back, and right than left." The left side was associated with coldness (as well as weakness), with the obvious conflict being that the heart is true to the left side. Aristotle was quick to say that this discrepant placement was only for the purpose of bringing heat to an otherwise cooler side of the body. In fact, the theory of hot and cold sides of the womb produc- ing boys or girls (respectively) persisted throughout Greek medical thought as much as anything that reflected left- right suppositions (these theories of "visceral distribution" would again become popular in the nineteenth century).

The teachings of Aristotle were passed down after the dawn of Christ and into the schooling of Galen, who was one of the last scholars of antiquity to contemplate these subjects.

*   *   *

The influence of doctrine is worth considering when con- templating how we became vastly distorted in our physi- cal makeup. Why we have lived, and continue to live, in a world so suspiciously dominated by the right? In the 1909 essay, *La prééminence de la main droite*, or as translated, *The Preeminence of the Right Hand*, sociologist Robert Hertz attacks the idea of our handedness being a biological or anatomical predisposition, and evidences socio-cultural effects instead:

> *The preponderance of the right hand is obligatory, imposed by coercion, and guaranteed by sanctions; contrarily, a veritable prohibition weighs on the left hand and paralyzes it. The difference in value and function between the two sides of our body possess therefore in an extreme degree the characteristics of a social institution; and a study which tries to account for it belongs to sociology.*
> ROBERT HERTZ

He goes on to say that it is specifically the spirituality and religiosity of man—and the accompanying conventions implied by these beliefs—that have done the work of biasing the body. No other forces seem so strong in our daily actions than those which are guided by an apparent higher being.

Many religions of the world attempt to sketch a rubric to deal with the dualism with which we surround ourselves, or which we might say is inherently built into our surroundings. The religions of the West approach the problem as a polarity of good versus evil, and right versus wrong, whereas philosophies from the East attempt to reconcile the opposing forces of the universe, finding harmony in the middle. You would never hear a Christian speak of morning prayer to God and afternoon mediation with Satan; however, a Buddhist would find a meeting of the two both inevitable and necessary in attaining balance and enlightenment.

The reconciliation of opposing forces is called The Middle Way for Buddhists. This was the concept passed down from Siddhartha Gautama (later known as Buddha), which stated that our most essential function should be to lead a life of moderation, balance, and neutrality. It is quite apparent that in attaining all of these virtues, one must know which extremes to look out for—so what are they?

Buddha set the stage for opposing pairs by suggesting that truth exists in the equilibrium of *eternalism* and *annihilationism*. This state of flux is the basis for the cyclical, rather than linear, flow of time to which Buddhists prescribe. Building on the teaching of Buddha, the sixth century Buddhist monk Huineng applied an addendum of thirty-six pairs of extremes that pervade our external world, our ineffable language, and our essential nature:

Sky and Earth
Sun and Moon
Light and Dark
Yin and Yang
Water and Fire
Words and Things
Being and Non-being
Physical and Non-physical
Perceptible and Imperceptible
Contaminated and Uncontaminated
Matter and Emptiness
Motion and Stillness
Purity and Pollution
Holy and Ordinary
Clergy and Lay
Old and Young
Great and Small
Strengths and Weaknesses
Perversion and Rectitude
Ignorance and Insight
Folly and Wisdom
Disorder and Stability
Kindness and Viciousness
Morality and Wrongdoing
Honesty and Crookedness
Truth and Falsehood
Fairness and Bias
Affliction and Malevolence
Permanence and Impermanence
Compassion and Malevolence

> Delight and Anger
> Generosity and Stinginess
> Progress and Regression
> Origination and Destruction
> Reality and the Physical Body
> Projection Body and Reward Body

Buddhist philosophy is cognizant of our dual nature; however, it teaches that the ultimate goal is to transcend this realm. The doctrines and practices laid out in the ancient Buddhist handbook *The Awakening of Faith* make it clear that Buddhism is, in fact, a motion against duality, because its end goal is oneness. One strives to become "one" with the world during meditation, not "two." This point was recognized by Carl Jung, the founder of analytical psychology, who spent many of his years making sense of this Eastern tradition. Although Jung recognized the concept of opposites, he was one of the first to methodically propose Descartes' error as a reflection of Eastern ideals: the mind and body are one and the same, not separate entities. Jung thought that when someone reconciled the conscious and unconscious—the self and the Self—they would be achieving Nirvana in much the same way as Buddhists do through The Middle Way.

Some philosophies do not preach of the transcendence that is inherent in both Jung's teachings and Buddhism, and rather call simply for a moderation of opposing forces, a concept better known as the Golden Mean. The tradition of balance goes back as far as records exist. One of three phrases carved into the temple at Delphi is *Mēdén Ágan*, which translates as, "Nothing in excess." A special cup that is credited to Pythagoras, known as a Tantalus or Greedy Cup, utilizes a series of intricate vessels to expel the entire contents of the cup if it is filled with too much liquid (presumably alcohol!). Socrates said a man "must know how to choose the mean and avoid the extremes on either side." Aristotle speaks of virtues in terms of their middle states, and offers that men and women are often unconsciously bound to one or another extreme imposed on them by society. Jews restrict their palette to kosher items, Christians often relieve themselves of a vice for the duration of Lent, and nearly every religion from Mormonism to Paganism is familiar with fasting. Indeed, moderate Muslims often approach the Quran as a book of moral relativism, middling its teachings to what is socially acceptable.

Any mention of "moderation" in philosophy or religion is a tell-tale sign that there is an imposing element of dualism. In fact, some occults and tribes that practice mysticism rebel against the notion of moderation and middle ground, and go instead by the way of the Left-hand Path, immersing themselves in taboo, and abandoning all adherence to established morals or ethics.

You may have noticed in Huineng's table of opposites that Yin and Yang appear midway through; however, they are the foremost recognized dualities in all of Eastern culture. This symbology has been most pervasive in the teachings of Taoism (Tao means "the way" or "the path"), which remains the reigning doctrine of ethics and philosophy that most Chinese follow today. The iconic symbol flows from black into white, with each section having the presence of the other color within it. Black and white are intertwined, interlocked, and inseparable, with their spirality signifying a world that is ever-changing and impermanent. Yin-yang dualism is perhaps the most long-standing and widely recognized representation of the human propensity to dichotomize. As historian Anne Harrington says, we have always had the tendency of "pitting reason against intuition, science against art" and of course, "yang against yin."

Yin-Yang Symbol

Within the yin-yang iconography, deep meaning is paired with an asymmetrical disposition, which introduces the pervasive nature of left and right in Eastern dualism. Yin (the black side with a white dot) embodies the female form, the Earth, things orientated to the back and right, and even numbers. Yang (the white side with a black dot) is opposite in every way, representing masculinity, the sky, the front and left, and odd numbers. Neither Yin nor Yang is regarded as supreme, meaning that right-handers are no more sacred, and left-handers are no more profane, which stands in opposition with many other traditions.

At large, the etiquette of Chinese culture can be swayed between the complementary components of Yin and Yang, depending on the era or ruler. The dynasty of Yu the Great reigned under the sign of the Earth, and hence was dominated by Yin and the right. The story is told that he had such a strong body preference that his left foot always dragged behind, never advancing before the right. T'ang the Victorious took reign directly after Yu the Great and ruled under the rising sun, which subsequently imposed Yang on his people. Some myths go as far as characterizing these rulers as fully hemiplegic, born paralyzed of one half.

The Hindu religion adds quite beautifully to the Oriental traditions that speak to common dualisms among us: good and evil, male and female, right and left. The following excerpt is a hymn from an ancient Hindu stotra outlining not only two beings in one, but one being in two:

> *Her sparkling ear ring is studded with blazing precious stones and he is adorned with a terrifying snake as his ear ring. Supreme auspiciousness is the essence of both Sri Shiva and Lord Shiva. Such is the wonderful form of Ardhanarishvara.*
> ARDHANARISHVARA STOTRA

Ardhanarishvara is the androgyny of the feminine, sparkling, radiant Sri Shiva, and the masculine, commanding Lord Shiva. Of the many stories applied to Ardhanarishvara, each addresses a common conundrum: if all gods are men, where do babies come from? Other mythologies follow a similar template and archetype, and depending on your viewpoint, they have either succumbed to or transcended dualism by making their deities either genderless or hermaphroditic. This was exactly the predicament that anthropologist Claude Lévi-Strauss blamed for our obsession with dualism. In his vast survey of mythological motifs, he found that stories of dual nature were always resolved by a "mediating term." In Hinduism this mediation is more than once found in androgynous gods, and we will soon see that even the story of our brains has a middling entity.

The holy magistrates that were in charge of introducing a female into male-centered theology had an important decision to make: what would this god look like? More specifically, how are the male and female attributes distributed? How about, down the middle? But then, which side is the male, and which the female? The Sanskrit that underlies the stotra makes a very clear distinction in the subtlety of its translation that the studded earring of the female form is on the left side of the body, and the snake earring of the male is on the right, which is how all stories of Ardhanarishvara have persisted throughout their many versions, both verbal and written. This choice could have been made with no more than a coin toss, but it goes well beyond chance that this and many other stories have been passed down in accordance with this particular asymmetry.

Judaism has also found itself to one side in many of the prescribed rituals found in the Torah. For instance, one is told to dress from the right to the left, and to undress from the left to the right. The right hand should precede the left in putting on a shirt, pants, socks, and shoes. However, an exception is made when tying anything because of the paradoxical obligation made by Jews to tefillin, by which a box full of Torah Scriptures is bound to the left arm, allowing it to sit snugly against the heart. Tying anything is made counterintuitive; for instance, shoes must first be put on from right to left, but bound from left to right, because tying must always proceed from the left. As if this were not enough to consider, the word "right" itself is relative, and is really just a synonym for whichever hand you prefer to use. As we come to find out, Jews are not the only people who might prefer to have two right hands.

Christianity is equally expressive in its adulation for the right (and specifically, the right hand). It is undeniably clear in Matthew's Vision of Judgement:

All the nations will be gathered before Him,
and He will separate them one from another, as a
shepherd divides his sheep from the goats. And He
will set the sheep on His right hand, but the goats
on the left. Then the King will say to those on His
right hand, 'Come, you blessed of My Father, inherit
the kingdom prepared for you from the foundation of
the world...'

MATTHEW, CHAPTER 25, VERSES 32-34

This passage goes on to honor the sheep of the right hand—which were more valuable and obedient—and then curses the goats of the left hand "into everlasting fire, prepared for the devil and his angels." It seems that this verse inspired Michelangelo to use the right hand of Christ in *The Last Judgment*, angling it toward his divine imagery of the heavens, while the left hand is positioned downward in the direction of hell and misery. He also gave the right hand the job of creating man in his magnificent fresco *Creation of Adam*, a decision supported by the book of Romans: "Christ Jesus is the one who died—more than that, who was raised—who is at the right hand of God..." In Michelangelo's *Separation of Light from Darkness*, it is the right hand that splits day from night. Ironically, Michelangelo himself was a lefty!

The Christian symbolism in art is often subtle, yet far-reaching, and I have every intention of encouraging you to survey the hands used in religious imagery from now on! In William Blake's etching, *God Judging Adam*, it is God's left hand that holds a casting scepter in the copper relief, so that during the printing process, it was his right hand that held the power of judgment. A stone sculpture on the side of Notre-Dame Cathedral in Paris depicts the temptation of Adam and Eve, holding each other with their right hands and divided by the sacred tree and evil serpent, with the forbidden fruit held to Eve's mouth by her left hand. In much the same way, the sixteenth century Italian artist Tiziano Vecelli painted his *Adam and Eve* with the left hand of Eve reaching awkwardly across the front of her body in order to pluck the fruit from the tree. The famed polyglot Albrecht Dürer follows suit in his version of *Adam and Eve*, depicting Eve's left hand as the one that clenches the fruit. Many of the associations of Eve with the left side of the body are derived from traditions that claim the rib taken from Adam to create Eve was from his left side, although this isn't directly substantiated by Scripture.

Scripture and art have both played roles in the insinuation that the devil is left-handed, and the many nefarious words that translate directly into "left"—including the word "sinister"—make this connection clear and everpresent. There are very few places in the Bible that even speak of left-handedness. Possibly the first left-hander put into writing is found in the book of Judges. Ehud, the son of Gera, wore a sword on his right side, and drew it with his left hand. In the same book, there is also the story of the Israelites punishing the Benjamites, in which each side gathers massive armies to go to war over a tribal dispute. Three versions of the same line tell an interesting story about a select group of Benjamite soldiers:

> **NIV** Among all these soldiers there were seven hundred who were left-handed, each of whom could sling a stone at a hair and not miss.

> **New King James Version** Among all his people were seven hundred select men who were left-handed; every one could sling a stone at a hair's breadth and not miss.

The Message There were another seven hundred super-marksmen who were ambidextrous—they could sling a stone at a hair and not miss.

Most interesting is that the The Message—a version tailored for sermons—removes the mention of left-hand-edness altogether, supplanting it with ambidexterity. These excerpts rival most notions of the sinistral side in other biblical passages, bestowing it with honor and great dexterity.

Islam is the last religion in the Abrahamic tridactyl that also claims allegiance to the right, and as it is a chronological successor to Judaism and Christianity, this comes without much surprise. One is to eat and drink with the right hand, and to wash and hold the genitals with the left. To act in opposition with these mandates would be embracing the manners of Satan himself. It has been recorded that Allah is a god made of two right hands, with "nothing left-handed about him"—both sides are known as Yameen in the native tongue, meaning "of the right side." An alternative translation of Yameen is "oath," which ties the right hand with the making of promises and contracts, practices that are done (presumably) under divine witness. The Black Stone inside of Kaaba at Mecca is called the "Right hand of Allah," and is the hand that connects the holy spirit to his people on Earth. It remains a mystery why Tawaf, the circumambulatory praise of Kaaba, is done counter-clockwise, which places the left hand of the pilgrims toward the Sacred House instead of the right; although many explanations exist relating to geographic coordinates, none are withstanding.

Could it be possible that not only the origins, but also the maintenance of our pre-eminent right-handedness is found in religion? If so, that means that tradition and etiquette are forces quite stronger than we might give them credit for, and persist across much larger scales than we could fathom. This would be a proof of how extraordinarily powerful small perturbations in societies can alter the history of the world, as well as our own physical makeup.

Imagine that you are cruising down the road in a large truck and you come to an intersection that has been blocked by a uniformed cop waving his hand in a slow circular motion. You glance toward the direction in which he is waving and realize he is directing you to the left for a detour onto a cross street. Your approach to this scene is quite predictable: you follow the detour. What was it that just shaped your decision to drive your massive automobile in a completely different direction than you were planning on going? Mere social forces under the right conditions are harborers of massive sway.

It is completely possible that the beginning of handedness was relatively unassertive, but from a small seed planted by society—perhaps from within these holy sects and doctrines—a revolution in our views on handedness formed. It set the stage for the contemplation of right and left, weaving into the worlds philosophies, and surely all further discourse on the matter. Each hand was its own being, with its own character and temperaments, and its own connections to divine and sinister powers.

*   *   *

"Middle Egypt" is home to an ancient Egyptian burial ground called Beni Hasan, abounding with tombs constructed as far back as 2100 B.C. Inside, there are paintings, inscriptions, and engravings that mostly tell the life of one man who had obvious influence over the land, the people, and perhaps, the local gods. Some of the stories were concerned with his daily activities and the manual labor done on the land, some told of great threats and conflicts, and others related to the tracking and hunting of prey. Most certainly, the Egyptian penman creating these biographical murals had no idea that they would be implicated in the discussion of handedness thousands of years later. The actions in the etchings—provided they accurately represent the populous of the time—give important clues to how people moved, and by which side of their bodies these movements took place.

Handedness from the relics of Beni Hasan was recorded in the late nineteenth century and was soon after made available to researchers. In a study on handedness by Wayne Brennan at Brooklyn College in 1958, all of the images that depicted humans performing an action which could be associated with hand dominance were tabulated. Casual actions or poses were negated, with the following results:

Actions	Left	Right
Using a bow and arrow	1	31
Using a battle axe	3	12
Using a knife	1	12
Making or sharpening a knife	0	3
Writing	2	8
Using a whip	0	9
Sickling grain	0	9
Painting or drawing	0	4
Stirring	0	4
Miscellaneous	2	13
Totals	9	111

The final tally, coming in at 92.5% right-handed, is in astonishing agreement with modern statistics on handedness, making this study eye-catching, but not without raising some follow-up questions. Did the Egyptians choose prototype caricatures, which were, for some other reason, predominantly right-handed, to appear in their drawings? Was there one artist who, at random, choose the right hand for actions and whose work was copied far into the future? Is this statistic biased by the linear arrays of men performing similar actions with similar hands? Or, is it wisest to choose the explanation which takes the shortest leap of the imagination, to side with Occam and his razor: we have been right-handed for a very long time.

One study analyzed the depictions of people in over 1,100 pieces of artwork spanning from the Middle Ages through the 1950's, and came to nearly the exact same statistical conclusion. This evidences more than just religious motifs in recording the bias of right-handedness: the right hand holds the pitch fork in Grant Wood's *American Gothica*, protects the ermine in Leonardo Da Vinci's *The Lady With the Ermine*, and stabs Peter in Paul Rubens' *Massacre of the Innocent*.

Other evidence comes from some of the first drawings ever recovered, presumably created by an ancestral primate. Many readers may be familiar with the activity in which school children trace their hands and then decorate the tracings to look like turkeys when Thanksgiving comes around. This is not too far from what was done nearly 50,000 years ago by our ancestor, the Cro-Magnon, in the caves of Chauvet, France. Although the Cro-Magnon tracings were void of cute decorations, nearly 80% of the hand outlines found were traced around a left hand. Once again, unless there is something more mysterious at play, the simplest explanation is that the right hand has long been the preferred hand for dexterity.

Altogether, this puts the decisive turning point—the dawn of handedness—beyond and before organized civilization and formalized culture. The drawings, the paintings, the etchings, the scribbles and scrawls, the Scriptures and poems, the art, the literature, the dossiers and documents, the deeds and indentures, the covenants of recorded past, and everything recoverable that we attribute to our species, has been fostered and fashioned by the right.

*　　*　　*

Many societies, clans, nations, and groups have made the distinction between right and left. The dipoles have reached much farther than Pythagoras could likely have imagined at the time of his sustoichia. Anthropologists have not overlooked the matter, and here I present an unavoidably abbreviated list of the predominant relationships and associations found around the world which have an antithesis in one another, from the Meru of Kenya, to the Toradja of central Celebes:

Right	Left
North	South
White Clans	Black Clans
Day	Night
First Wife	Co-Wife
Senior	Junior
Man	Woman/Child
Superior	Inferior
East	West
Sunrise	Sunset

Sun	Moon
Light	Darkness
Political Power	Religious Authority
Successors	Predecessors
Older	Younger
White Man	Black Man
Cultivation	Honey-Collection
Hand Used for Eating	Hand Used In Defecation
Hand Used for Completing Payments	Hand Used for Initiating Payments
Upper Hand	Lower Hand
Female Hand for Sex Play	Male Hand for Sex Play
Masculine	Feminine
First	Second
Ahead	Behind
Above	Below
Upstream	Downstream
Strong	Weak
Settlement	Bush
Regular or Stable (Life?)	Difficult to Control Death
Auspicious Omens	Inauspicious Omens
Important Omens	Unimportant Omens
Sperm	Blood of Conception
Bones, Teeth, Hair	Blood of Body
Solid	Fluid
Land	Water and Rain
Mountain or Hill	Valley or Sea
White	Red
Normal or Cool	Warm or Hot
Ideally Active	Ideally Passive
Father's Matriclan	Mother's Matriclan

Formal or Respectful Relations	Joking and Sexual Relations
Purity	Pollution
First Group In Ritual	Second Group In Ritual
Center-Pole	Hearthstones
Cooked	Uncooked
Initiated	Uninitiated
Heaven or Sky	Earth
Spiritual	Worldly
Interior	Exterior
In Front	Behind
Old	New
Good	Evil
Life	Death
Ancestral Spirits	Evil Spirits
Shaman	Sorcerer
Afterworld	Underworld
Abundance	Poverty
Fullness	Hunger
Side Men Are Buried	Side Women Are Buried
Bow	Drum
Bush-Clearing	Seed Planting
Normal, Esteemed	Hated
Brewing	Cooking
King	Queen
Chief	Subject
Owner Of Land	Hunter
Health	Sickness
Joy	Sorrow
Wealth	Poverty
Security	Danger

Pure	Impure
Even	Odd
Hard	Soft
Princess	Diviner
Political Rank	Mystical Office
Legitimacy	Illegitimacy
Normal Birth	Twin Birth
Cattle	Chickens, Sheep
Milking	Hunting
Clothed	Naked
Shaven Hair	Long Hair
Barkcloth	Animal Skins
Civilization	Savagery
Royal Endogamy	Misalliance
Fidelity	Adultery
Personal Combat	Murder
Culture	Nature
Classified	Anomalous
Order	Disorder
Ordinary Men	Witches And Sorcerers
Persons	Things
Soul	Spirit
Elders	Prophets and Rainmakers
Oracles	Diviners
Shrines Inside Homestead	Shrines Outside Homestead
Ghosts	Spirits
Kinship and Marriage	Incest and Cannibalism
Wives	Leper Women
Ancestors	Heroes
Amateur Craftsman	Blacksmith
Ordinary Housewife	Pottress

Ordinary Blood	Menstrual Blood
Melted Iron	Iron Ore
Fired Clay	Raw Clay
Genealogy	Myth
Sexual Intercourse	Sexual Abstinence
Moral	Amoral
Authority	Power
Normality	Inversion
Moderation	Immoderation
Cosmos	Chaos
Completeness	Incompleteness

This list is quite indicative of the aura that has surrounded the right-left dichotomy and its couplings for
much of history. These dualities span societies and arise
in agreement in what appears to be bursts of spontaneity.
From the innards of densely-populated cities to the far-
thest-reaching and most remote establishments of scorch-
ing plains and humid jungles, human solidarity is most
evident in the predisposition to apply dual classification to
nearly everything.

As the simple words "right" and "left" have come to
be, under different regimes and through a chain of generations, they have absorbed definitions that suit them
based on a long list of imputations. Consider, however,
that meaning presumably came before the proper tooling
of coherent language. The distinction—the realization!—
of asymmetry was a necessary revolution, enlightenment,
and emancipation before these words ever had reason to
come about, and before the clashing of distinctions was
made. A strong sense of embodiment—being able to tell
one body part from another—may very well be the distin-
guishing element to look for in our species' history (and in

the history of other species, as well) to determine why we have veered so radically into a two-sided world. Are we the only creatures that understand that all of our body parts put together equal a whole? Is this why dogs chase their tails, and cats scare themselves in the mirror? When did we realize our left is not our right?

One of the first clues to finding the emergence of a dualistic self-awareness was brought about by the early twentieth-century psychologist and researcher, Jean Piaget. He was able to trace the development of children's' sense of embodiment through objective questioning. One conversation with a small boy of seven-and-a-half years old perfectly exhibits the reality of a child, and the challenge this presents for a scientist. Sitting across from the boy, Piaget sets down a pencil and a penny:

Is the penny to the right or on the left of the pencil?
The left.
And the pencil?
The right.

Next, Piaget asks the boy to switch positions with him, bringing him to the other side of the table.

And now is the penny on the right or on the left of the pencil?
The left.
Really?
Yes.
And the pencil?
The right.

JEAN PIAGET, THE ORIGINS OF INTELLIGENCE
IN CHILDREN

The boy's answers were the same, even as he rotated to the other side of the table. When asked how he had formulated his response, the boy said, "Easy, I remembered how they were before." Strong associations to the right and left fields of space (the ones we take for granted everyday) were shown to be still-forming over and again through this type of introspection.

Piaget revolutionized the way that we think about childhood development, which also provides clues to the actions and habituations of mammals with high, "childlike" cognitive abilities. He believed that children don't simply act as immature or less intelligible adults; rather, their approach is fundamentally different. For kids younger than seven, there is not objective quality about their reality, and they float about in an egocentric state. They will interrupt conversations, take candy from stores without paying, and treat someone else's house as if it were their own because they have no internal reference built up that distinguishes "mine" from "theirs." Kids this age can relay a story or set of instructions, but Piaget found that they don't really care if the person they are telling it to actually understands what they are saying. A child's misgivings are not coming from a lack of empathy, but instead a lack of reference. As children near puberty they begin to understand the difference between individual and social life, and following that, the difference between their right and left.

This is perhaps the root cause of the difficulty that most young children face when writing alphabet letters that have left-right symmetry, like "b" and "d." Although we do grow out of the most evident confusions, researchers can still find the relics of the perplexed animal inside us. In 1979, William Ferrell at Stanford University flashed variously orientated arrows (up, down, left, right) to subjects on a screen and asked them to call out the direction in which the arrows were pointing. It turned out that there was a marked increase in the time it took to discriminate left or right arrows than ones that were up or down. A plethora of evidence also finds that we have an incredibly easy time remembering objects we have seen before even if they are flipped about the right-left axis, and we gradually become worse as the axis of symmetry rotates. Most of the foundational connections in our brains are, after all, highly symmetrical. This means that our thoughts and perceptions are actually undergoing some repetition so that each hemisphere has the proper information to make decisions with. For instance, when a single hemisphere's visual cortex is electrically stimulated, phantom images known as *phosphenes* (minor hallucinations) appear in perfect symmetry to both the left and right eyes.

Every single animal, including fish, birds, rats, cats, dogs, and monkeys cannot (and never do) create strict associations between the left and right sides of their world, while some humans have made this primal trait something of a toy. Lewis Carroll was playful with his ability to write backwards, and he developed wonderful imagery and allusions to mirror worlds in his *Alice's Adventures in Wonderland*. Leonardo Da Vinci is likely the most famous mirror writer, using the technique many times to codify his work. If it were not for these and many other cases of normal individuals being able to read, write, and even speak in reverse, we might not be so convinced of the looking glass sitting between our two hemispheres. To some, the distinction of right versus left comes naturally and automatically, while to others, it remains completely inorganic and cumbersome. Richard Feynman is noted as having used a birthmark to remember which side was his right, Sigmund Freud is said to have penned a few strokes in the air in order to remind himself, and some Roman Catholics rely on signing the cross (a movement strictly bound to the right hand).

Embodiment can be examined through the evolution of language, and tells a gripping story of how "right" and "left" have taken on different meanings over time. The Indo-European languages consist of over 400 languages and dialects, including many of the most recognized and widely-used tongues in the world. They coalesced in the Mediterranean nearly four thousand years ago, and, as the early languages spread, they reshaped themselves and morphed into forms that still make it difficult—if not impossible—to retrace their precise journey and genealogy. The vast amounts of land this transmission covered makes it a particularly interesting etymological case study. Its major divisions are denoted by the Anatolian, Hellenic, Indo-Iranian, Italic, Celtic, Germanic, Armenian, Tocharian, Balto-Slavic, and Albanian subgroups. Modern English is one of the youngest languages contained within this collection, being only 500 years old.

The following is an inventory of documented translations, organized as closely to their shared etymology as is possible. If you read carefully, you can pick up on some subtle metamorphoses, and some abrupt shifts from one line to the next:

Language	Right	Left
Greek	δεξιός	αριστερός
Mycenaean	de-ki-si-wo	-
Latin	dexter	sinister, laevus, scaevus
Catalan	dret	esquerra
Italian	destro	sinistro

French	droit	gauche
Haitian Creole	dwa	gòch
Spanish	diestro, derecho	izquierdo, siniestro
Portuguese	direito	cnahoto
Galacian	dereito	esquerda
Romanian	dreapta	stânga
Irish	deas, ceart	clé
Welsh	de, deheu, hawl	aswy, chwith
Breton	dehou	kleiz
Gothic	taihswa	hleiduma
Icelandic	hægri	vinstri
Danish	højre	venstre
Swedish	höger	vänster
Norwegian	høyre	venstre
Old/Middle English	riht	lift, luft
English	right	left
Dutch	recht	linker
Afrikaans	reg	links
Old German	zeso, reht	winster, linc
German	recht	links
Lithuanian	teisė	į kairę
Lettic	labs	kreiss
Croatian/Serbian	u pravu	lijeva
Bosnian	pravo	lijevo
Slovak	právo	vľavo
Slovenian	pravica	levo
Czech	právo	vlevo
Polish	prawy	lewy
Russian	pravyj	levyj
Sanskrit	dakşina	savya, vāma
Avestan	daşina	haoya, vairyastāra

Tocharian A	pāci	śālyās
Tocharian B	śwālyai	saiwai
Hittite	kunna	GÙB-la
Akkadian	imnu, imittu	šumēlu
Ugaritic	ymn	(u)sm'al
Hebrew	yamin	semo'l
Arabic	yamîne	šimâl
Albanian	e drejtë	majtë
Armenian	aǰ	jax
Azerbaijani	sağ	sol

For nearly every word meaning "right," there is a strikingly positive semantical implication. In this family of languages, there are only three roots for the "right side," all indicating the "good" or more comfortable side. Juxtaposed to these are the roots for "left," which vary, but find their semantics replete with negativity, relating to that which is crooked, weak, and foolish.

The English word for "right" came from the Old English word "riht," which was used to describe something straight. "Left" similarly came from "lyft," which indicated weakness in structure or virtue. Latin uses the familiar word "dexter" for the right, which is also used today when describing the precision one has with his or her hands, as in the word "dexterous." Contrarily, the Latin word "sinister," representing the left, has taken on obvious negative semantics and bad connotations. Both Latin and French have undergone revisions to change the original word that meant "left" because it expressed such negativity.

In English, "right" can be an affirmation of correctness, something which is morally sound, and an unalienable endowment to members of society. Our "right hand man" is often that whom we trust and care about; a "left-hand compliment" is actually an insult; the right hand makes oaths while the left hand has historically delivered the *coup de grâce* to a dying adversary. These pairings begin to reveal the subtlety of any synonym that is to stem from the opposite sides. What follows is once again abbreviated, but shows some of the words you might have once used yourself that quite literally mean "right" and "left."

Right	Left
Ability	Awkward
Alliance	Bad
Charity	Bad Day
Comfortable	Bad Finances
Commitment	Bad Luck
Competency	Badly Done
Correct	Cleverness
Dextrous	Clumsiness
Entitlement	Confused
Entitlement	Crooked
Favorable	Damage
Fidelity	Defective
Fitting	Difficult
Good	Disaster
Grace	Evil
Health	Extramarital Relationship
High Ranked	Faulty
Honorable	Impolite
Human Rights	Lameness
Husband	Liar

Jurisprudence	Limp
Just	Lopsided
Justice	Lower Position
Luck	Maimed
Male	Mistaken
Mercy	Negative
North	Ominous
Positive Action	Paralysis
Promise	Perplexed
Proper	Ponderous
Right	Puzzled
South	Questionable
Straight	Sneaky
The Best	Tired
The Law	To Die
The Leader	To Disregard
To Please	To Neglect
Truth	Ungraceful
Valid	Untrustworthy
	Weak
	Wife
	With Hate
	Woman
	Worthless
	Wrong
	Wrong Direction
	Wrong Road

It is rare that in any language that "left" be sided with auspiciousness and good; however, it does occur in some Scandinavian languages such as Danish, Swedish,

Icelandic, and Norwegian. There are even scarce records—hints of this inversion—in very ancient Greek texts. Why is it that there are so few *good* words associated with "left," and in so few languages? The best explanation takes two logical steps. The first is the argument that this entire chapter has flirted with: that we are predisposed to applying opposing symbolic classifications to the world around us. The second necessary claim has to do with the power of majority. If the majority associates their way with *good*, and the majority is mostly right-handed, then it would follow that anything *good* should be "right." This is the basis of how homonyms are synthesized throughout society. This means that the record of humans stigmatizing, oppressing, and casting out minorities—left-handers in this case—is literally sewn into the words of our language, and has been for millennia.

Across planet Earth, the bias in vocabulary is nearly as strong as handedness itself. It is found in Albanian and Ambonese, Danish and Dutch, English and Eipo, French and Fulani, German and Gogo, Italian and Indonesian, Korean and Kaguru, Latvian and Lugbara, Meru and Mapuche, Nyoro and Norwegian, Russian and Romanian, and Spanish and Swahili. It is a phenomenon that tells a fantastically riveting story of who we are, how we came to be, and the indelible nature imbibed and injected into the life and times of the extraordinary social animal that we are.

CHAPTER FIVE

Split Brain

We have a sufficiently strong propensity not
only to make divisions in knowledge where there
are none in nature, and then to impose the divisions
on nature, making the reality thus conformable
to the idea, but to go further, and to convert the
generalizations made from observation into positive
entities, permitting for the future these artificial
creations to tyrannize over the understanding.
— HENRY MAUDSLEY

The nervous system, I repeat, is
physically double.
— J. HUGHLINGS JACKSON

Communication across the revolutionary divide
is inevitably partial.
— THOMAS KUHN

*   *   *

Surprisingly, we have not yet looked at the brain itself, and only so far considered it in a distant, historical context. It would be surprising if one didn't know how much the brain weighed; clichéd phrases that pair the words "three pounds" with words like "mass," or "matter," or even "jelly," are all too frequent. What most people do not know is how the brain works, and how this small organ—yes, of only about three pounds—facilitates both automatic and intentional actions of the body.

The brain consists of roughly eighty-six billion *neurons*, and while these are not necessarily the only building blocks of the organ, let's just say if you had to choose a type of cell to lose up there, your first pick would not be the neuron. In the most fundamental sense, neurons are just like any other cell, all sharing a common structure of an outer membrane, cytoplasm, nucleus, and particular regulatory processes that keep them alive. Where a neuron differs, from say, the plant cell you put under the microscope in grade school, is that the neuron is part of an incredibly vast network of other neurons, that when put together, are able to calculate, compute, decide, learn, and somehow come together to form the mind.

Typically, a neuron will send information through one large nerve fiber called an *axon*. The axon of a neuron can project to its nearest neighbor neuron or across extraordinarily large distances, some even spanning from inside the brain down the entire length of the spinal cord. Almost every nerve fiber like this is coated with a very thin sheath of an insulating material called *myelin*. This myelin not only protects the nerve fiber from damage, but it acts as an electrical insulator. If we didn't have myelin, every neuron would essentially become connected and entangled into one giant mess with its neighbors, and there would be no way for a signal to make it to its destination without shorting out. It is for this reason that any disease in the body which disrupts the proper application of myelin is going to cause major issues—loss of hearing, eyesight, and movement of both body and mind. The glossy white texture of nerve fibers can be attributed to myelin, sometimes referred to as the *white matter* of the brain. On the contrary, neurons themselves, which do not have any protective sheath (except for their thin membranes), are called the *gray matter*, because of their contrastingly darker appearance when compared to white matter. These two simplistic naming conventions are relics from the first anatomists who made the blunt distinction before even knowing what each type of matter actually did.

The final parts of a neuron are the tree-like branches that fan out from the cell, called *dendrites*. This makes the brain cell act like a big dream catcher, or fishing net, allowing it to latch onto the outgoing axons of other neurons, creating the "network" for which the brain is so famous. The connection points between two neurons are called *synapses*, and indeed, these too have a complex range of dynamics all their own.

The way in which the brain talks amongst itself is still a largely unsolved mystery, with more theories than could fill this book, but we do have a rudimentary understanding of the principles that underlie the way in which a neuron sends and receives signals. An *action potential* refers to the brief, yet large jump in a neuron's voltage, which forms the currency for all communication between brain cells. Often just called "spikes," there are two basic ways in which they are leveraged by our bodies. The first controls the automatic processes, such as making our heartbeat, our lungs inhale and exhale, and the muscles of our digestive system chug away. If you have ever placed your ear to someone's abdomen, you are familiar with the cacophony of noises— slurps, gurgles, and whines—hidden within. The neurons that are acting out these processes are, for the most part, located deep in our brains, where the spinal cord and brainstem merge to form the midbrain. This spike itself has particular characteristics concerning its timing in relation to other spikes, height, width, and so on that give it a unique character as it travels in the brain. In autonomic function, after the spike occurs, the neuron undergoes a systematic operation to reset itself and continue its job.

You can imagine that these automatic processes don't require intricate computation; otherwise we would have an incredibly inefficient brain, sucking up excessive power and attention. Try this: don't think about blinking. I repeat, whatever you do, *do not* think about blinking. It is almost guaranteed that you are now mildly frustrated, because even though I asked you to ignore something, it feels as though you can't shut your eyes automatically anymore. This phenomenon has been famously exposed by George Lakoff, and works just as well if I were to tell you to *not think of a pink elephant*. Lakoff calls this the *ironic process theory*, but the concept has existed in psychology since William James dubbed it the *dual process theory* and aimed to reconcile how one action can be controlled in two different ways. For the last few paragraphs prior to this one, you were also blinking, but do you remember those blinks?

As you became aware of blinking, you transferred control of that muscle reflex from an automatic process to a conscious one. This intentional brain signal is the second way in which an action potential can be generated. As signals from many areas of the brain collude onto the dendrites of a single neuron, they become provocateurs and can make the neuron fire. This is a calculated action of the brain, and one which we consciously control. If free will does exist, this is one of the processes by which we can study it.

Up to ten thousand axons can converge onto the branches of a single neuron, all providing dynamic input that will influence its decision making process. Those axons may exist on different branches from each other, at different distances from the neuron's cell body, and they may each use a different combination of *neurotransmitter* to induce their effects across the synapse. All of these variables play into the equation that decides whether or not a neuron will fire at a given instant. If enough of the inputs from the dendrites are telling the neuron to fire, then it will. However, the signal from one branch on the tree may be saying "fire," while the other branch is saying "be quiet," thereby putting the neuron in a very delicate political war, while it balances on an edge. This process has a nearly infinite number of conditions, all of which undergo extreme fine-tuning through *neural plasticity* both at the beginning of, and throughout, our lifetimes.

And that is just the neuron. We won't go further into the molecular and genetic interactions, although they compose some of the most fantastic topics of modern research, but we will expand outwards and consider how groups of neurons collaborate.

The brain can be thought of as containing various compartments of densely packed neurons, usually all with very similar characteristics. A group of alike neurons is called a *nuclei*, and although each neuron still has a very important job in deciding when it will fire, in many cases, nuclei will make various computations within themselves, only to relay their final decisions elsewhere through fibrous white matter. The nuclei are the subcommittees of neuronal processing. They may be in charge of relaying sensation from your fingertips to the sensory areas of your brain, they may control the movement of your legs, and they may serve as entirely internal processors that figure out exactly how to react to the changing of a red light to green at an intersection.

Sometimes, because these nuclei do act independently, they also perish with each other. This is the case in Parkinson's disease, in which an entire group of neurons containing the neurotransmitter *dopamine* dies away. The disease often leaves people unable to walk correctly, with uncontrollable tremors, or completely lifeless altogether, trapped in a jail cell of a body that can breathe and think, but not express. We will all face the fact that our brain cells do not have infinite lifetimes as our bodies begin to out-live our minds. However, knowing that they often expire together as groups of nuclei, is the first step towards creat-ing targeted treatments for very specific pathologies. For instance, modern technologies are making it possible to graft brand-new cells into the dying nucleus of a Parkin-son's patient, and alternative treatments exist that rely on implantable electronic hardware programmed by a com-puter to regain control of lost abilities. The medical profes-sional's toolbox used in treating neurological dysfunction is rapidly extending far beyond pills and serums into cut-ting edge genetics, nano-sized robotics, and optically con-trolled brain centers.

One cubic millimeter of brain tissue contains nearly half a billion brain cells, and at this size, it becomes just large enough to see with the human eye. Yet, there are mys-teries and complexities that we are still struggling to grasp on this scale—presuming they are indeed graspable. Let me ask you, would it be possible for you to count to eighty-six billion? I am sure the answer is yes, given enough time, of course. What this demonstrates is simply that there is not a single neuron for every number that we are capable of pro-ducing or every memory that we have. Rather, it is small ensembles of neurons and nuclei that enable a higher-order emergence of ability, knowledge, and awareness. If it were any other way, we would certainly be limited to speaking through pre-made sentences and moving in well-defined directions—we would be as mechanical as a Venus flytrap. These groups of neurons are the seats of our intelligence.

Anatomical variation in the brain begins to occur on a large scale when groups of neurons take shape, segregated by tracts of white matter or pockets of either fat or fluid. One of these encapsulated structures is the *cerebellum*, which sits just above and behind the neck. This structure is about the size of a closed fist, and contains more than half of the neurons in your head. It is in charge of coordinating muscle movement and maintaining equilibrium in the body. The brain that most people think of, though, is the larger curlicued ellipsoid, the *cerebrum*, which sits above the brain stem, and in front of the cerebellum. Here are the deep and mysterious brain structures, such as the thalamus, pituitary gland, amygdala, and hippocampus, which are distinctly in charge of everything from movement to sexual urges, and from fear processing to memory. Moving closer to the skull, we enter the area that is most exciting for us, and newest to our species, the *cerebral cortex*. It is that squishy-looking top to our imperfectly round brains, where neurons are packed in at the highest densities. We can attribute nearly everything that separates us from a monkey to the cortex. This is where we plan, strategize, and likely where consciousness (whatever that actually is) is housed.

Of course, this description of our anatomy would not be complete if it did not include the quite pertinent fact that starting as low as the brainstem, most of the structures in our brains are separated down the midline, creating, in fact, a multitude of hemispheres. This split may be the most important separation of two entities anywhere on Earth. It is sure to shape our perceptions and our internal calculations, and to alter the way in which we approach and reason with the world around us. If we did not have this separation, the entire concept of dualism itself may be distant and utterly obscure to us. How these two sides of the brain operate both together and independently is what we are about to explore, first in answering how they connect and function, and later, considering what it all means. We have now formally begun our journey into the brain.

* * *

When Socrates was on trial for blasphemy against the empowered religious zealots, he blamed his precision of thought and scrupulousness on his inner *daimon*, which acted as a voice of reason to nearly all his worldly concerns. It was the entity that warned him of rhetorical rabbit holes and guided him in the way of pure and good philosophical argument. He was neither the first nor the last to speak of the voice within. Years later, Adam Smith conceived it as an internal witness, and C. S. Lewis recognized the phenomenon as *consciousness*, and bound it to religion by saying it "is either inexplicable illusion, or else revelation."

This concept is rarely found without inalterable ties to the divine, being ever-more true for the eminent father of Western philosophy, René Descartes. He was a man nearly obsessed with the awareness of ourselves, and spent most of his life trying to formulate where this voice came from, and what it was doing there. That famous utterance, "I think, therefore I am," found in Descartes' writings in both French (*Je pens done je sues*) and more famously in Latin (*Cogito ergo sum*), was a statement about the human condition and mind-body dualism. The statement says something very powerful and implies that our soul, which is presumably in charge of our thinking (*res cogitans*), is a wholly separate entity from our body (*res extensa*).

Cartesian Dualism is surely one of the centerpieces in any contemplation of the split between body and soul, and Descartes is the reason why the words "mind" and "matter" sound so familiar together. In his view, the utterances from your mouth were the corporeal expression of the contemplations of your mind—there was truly a ghost within the machine.

Descartes' philosophy was still with holes unless he could reconcile where the singularized soul met up with the body proper. The brain, with its two large cerebral hemispheres, posed a major issue for him. By the sixteenth century, anatomy had placed just about everything worth philosophical contemplation above the neck, yet very little was known about the function of our massively entangled brain structures. For Descartes, there had to be a single entity that mediated for the soul and that connected us with the higher order of the cosmos. It was a great conundrum that led him to being one of the first philosophers to consider how symmetrical cranial organs seemed to emerge into a single conscious being.

In the midst of his studies of animal anatomy, while attempting to decipher the location of a unitary organ capable of housing the soul, Descartes seemed to strike gold: the pineal gland! This is a structure placed deep within the brain, nearly perfectly round and centered on the midplane, and without any outwardly noticeable signs of two sides. Long before Descartes, Galen wrote that this was the structure by which thoughts flowed in and out of the body, and so there were thorough a priori convictions in place as to the sacrosanct status of this tiny edifice. Could this be the epicenter that bridged the gap between the physical realm where the body existed, and the metaphysical realm where the mind was maintained?

Unfortunately for Descartes, he was one thin slice of membrane away from the disappointing truth that the pineal gland is also doubled in structure. Descartes came to face harsh criticism for his rather lackluster placement of the soul, but the mind-body problem remained, and he was hardly the only one interested in searching for the source of it.

A century after Descartes, Giovanni Lancisi, the man responsible for discovering how malaria was spread, also took to the insurmountable task of "seating the soul." Lancisi firmly pitted himself against the grain of those before him who maintained that the white matter in the brain was simply supporting tissue, or scaffolding, that kept everything from jostling around. He took a very close look at the *corpus callosum*, the largest white matter tract connecting the two hemispheres of the brain, and thought that it would be a perfect antenna for communicating with the transcendent. It met nearly every standard for that almighty component of the human body by which Descartes had meticulously guided his search, and most importantly of all, there was truly only one corpus callosum.

Whether or not it was Lancisi who refocused the medical community onto the corpus callosum is arguable, but it was certainly around his time, and within his discipline, that talk arose of the potential uses to the mind that a bridged structure like the corpus callosum could provide. It seemed as though mention of the soul quieted and was replaced with anatomically and psychologically deducible problems, like the possibility of double consciousness.

Detailed drawings and open anatomy sessions made it ever more apparent that inside us all were two grossly distinct brain hemispheres connected by one major bundle of nerve fibers. One could only be curious as to whether or not the two brains had separate personalities, or if they were just mirrored entities, operating as one and the same.

The very idea that each side of the brain was not simply a reflection of the other bothered many medical men of the nineteenth century, who were at that time preoccupied with the problem of coexistence, or the idea that there could be two minds within one person. Stories of single hemisphere hypnosis surged through popular gossip, and even attempted to provide proof of brain localization by turning off the left hemisphere, subsequently affecting Broca's Area and disabling language altogether. One critic, perhaps fed up with the insistence on dualism, would add that we "are not bound to the number two in considering the mass of conscious, subconscious, and unconscious states." However, the form of our anatomy had a strikingly visual rebuttal to this statement.

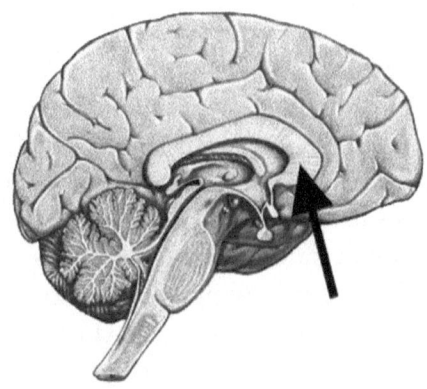

Corpus Callosum

As hypotheses of our dual nature were becoming formalized scientific endeavors, experimental psychologist Gustav Fechner began to ponder the hypothetical outcomes of splitting a man in half by severing the corpus callosum directly down the middle. He suggested that in such an event we would "undoubtedly see something equivalent to the doubling of a man." Fechner believed that each hemisphere of the brain contained the same dispositions, knowledge, memories, and the same consciousness in general. In line with his strict stance on a completely mirrored body and soul, Fechner went on to imagine that after being split, each half of the man would act independently and begin to live in separate domains, being developed by their environments accordingly. We will soon find out that he was indeed onto something.

Hardly anyone found the corpus callosum to be a trivial component as the nineteenth century passed, but it was still a vexing mystery as to what it did, what it withheld, and what would happen if it were to be tampered with.

* * *

"When you read this letter, I will be dead." This was the archetypal introduction left in a note written by forty-two-year-old Dr. Justine Sergent just before she and her husband committed suicide on April 10, 1994. They were found the day after in the garage of their Montreal duplex, sitting side-by-side in their car with a hose running from the exhaust outlet into a side window. Subsequently, the official coroner's report cited carbon dioxide poisoning as the cause of death; however, the two knives reportedly found on their dashboard were morbidly suggestive of their willingness to take extreme measures, had they needed to.

Sergent was an all-star in the scientific community. Born in Lebanon, she would later marry in France, and then move to Canada where she earned her Bachelor's, Master's, and Doctorate degrees in psychology at McGill University in Montreal. Her articles appeared in journals reserved for only the most groundbreaking work—*Nature*, *Science*, and *Brain*, to name a few. Despite being heralded by many communities for her scientific contributions, she was known within the university to be deliberately provocative, occasionally with the full intention of getting a rise out of her peers. It was surely this that in 1992 pushed an anonymous writer to send a letter to the *Montreal Gazette*, as well as to surrounding institutions (including Sergent's own McGill University) that sternly accused Sergent of scientific misconduct. At the heart of the matter was what Sergent would later call merely a "technical dispute, and only that," having to do with whether or not she was granted permission to employ a particular type of imaging on subjects in one of her studies.

Action was taken on Sergent by McGill shortly after the letter was exposed, reprimanding her and dissuading her from further work that involved conducting experiments without proper ethics-committee approval. In what should have been a purely internal procedural matter for McGill University, the *Montreal Gazette* took the lead on an investigation and publicly chastised Sergent for wasting public funds. As often is the case, the media became an overbearing force in Sergent's arbitration. It didn't help that just a few years earlier, Dr. Roger Poisson—also local to the Montreal area—was convicted of fraud, with his studies from over 1,500 patients and spanning over twelve years being thrown out of the academic arena. The bad taste that this left in the scientific communities' mouths put a persistent spotlight on Sergent's case. The day before her suicide, the *Gazette* ran a large article titled "Researcher Disciplined by McGill for Breaking Rules," and made mention of the connection between Dr. Sergent and Dr. Poisson's misgivings. To Sergent, it was as if her entire life's work and reputation had been sunk to the depths of a sea in which no working scientist wants to wade. In her own somber words that lie just above her signature—"Justine Sergent"—on her suicide note, she left the world with this:

*I had a rich and intense life, but there comes a
point when one can no longer fight and one needs
a rest. It is this rest that my husband, who has
supported me in all aspects of my activities and my
life, and myself have decided to take.*

JUSTINE SERGENT

As a colleague would note, Sergent was a unique investigator with an "unfinished opus." Her opus was one that dug deep into the functional differences between the two hemispheres of the brain—she was very obviously concerned with how they worked, and how they came together as an emergent marriage of the mind. She masterfully approached one of the fundamental questions in the field of brain laterality, pondering not what the two hemispheres do, but what one hemisphere does. Her experience with ablated patients—those who had lateral lesions, or hemispheric-specific surgeries—made her uniquely suited to provide answers.

The isolationist approach is important because if we can understand how one side of the brain works in a way that makes sense with respect to the whole, then the brain is more like a redundant, functionally equivalent, bisymmetric organ (like your eyes). However, the more interesting question arises if the contrary is true, and the sum is not simply equal to the constituent parts. When they are together, does something spectacular emerge?

There are three populations that are uniquely suited for the study of single-hemispheric function. The first group is born void of these important connections between each side of their brains, which includes the corpus callosum. This congenital malformation is otherwise known as *callosal agenesis* and is sometimes referred to as the "natural model" of the split-brain. The character that Dustin

Hoffman plays in the 1988 film *Rain Man*—a sensitive, yet brilliant autistic idiot savant—was based on a real man, Kim Peek, who was born with callosal agenesis. Although he suffered from many cognitive deficits, Peek was quick with calculations, had memorized every zip code in the United States, and was most well-known for his reading skills. He read books by scanning the left page with his left eye and the right page with his right eye. If you were able to read as fast as Peek, reading this book would take you about thirty minutes total, and moreover, you would retain about 98% of it to indefinite memory.

The second group are those who have undergone a procedure called a *hemispherectomy*, which entails removing an entire hemisphere of the brain. This could be in response to insurmountable deformation, or uncontrollable seizures as seen in epileptics. It is a procedure that is usually only suitable for children, as they are able to rely on adaption of the remaining hemisphere to control the most important executive functions in the brain. There is a particular morbidness in removing an entire half of someone's brain, but this has been heralded by many as a life-saving procedure, instilling some normality back into the lives of both the patient and their parents.

Removing half a brain during one of the most developmentally important times in one's life gives a good indication about what processes are still settling, and which ones have already been established. Many of the children are able to recover with only slight cognitive deficits but usually display some amount of paralysis to the limbs that were controlled by the removed brain hemisphere. Most interestingly, kids who have their left brain removed are usually able to regain control over speech and language, which perhaps points to an inhibiting role of the left hemisphere onto the right—removing the left might somehow be releasing the lexical shackles that have bound the right. However, it has also been shown that if the right hemisphere is forced to take on language, it can only do so to a certain extent, indicating that it is not an ideal substrate for the task.

The final set of patients that can shine light onto our two brains are those that have had some or all of their corpus callosum surgically severed in an operation called a *callosotomy*. Again, the most common reason for this operation is to cordon off the neuronal hyperactivity associated with epilepsy, and to prevent it from spreading between each hemisphere. It is much like thwarting a large-scale power outage by disconnecting a surge at a substation and not letting it pass through to other parts of the grid. Those who undergo surgery are often past the critical age in which neural plasticity can be effective. They do not have the opportunity that Peek did, or that children do to develop special skills in the absence of interhemispheric commissures. The functions that are removed are more or less a permanent deletion. For these reasons, they are most intriguing since what remains are two fully separated structures that have each had a wealth of experience, memories, and dealings with the world. We should therefore be interested in how they—and their two halves—reconcile this sectioning of the self.

* * *

Throughout history, the shadows that loom from the tragedies of war often blossom into profound medical opportunities. The characters in the story so-far-told of the brain are no exception and are indeed an exquisite example. Galen's insights as physician to the gladiators in Pergamon sowed the seeds for modern anatomy and philosophies of healing.

As the wrath of a saber met the pitch of a musket ball, Robert Knox was summoned to the Battle of Waterloo, where he became familiar with surgery, including rare amputation techniques, an acclamation that surely made dissections of resurrected corpses later in life so palatable. Cesare Lombroso began his long career of measuring heads and body parts as a military doctor in the wars for Italian unification. B. F. Skinner brought pigeons into the laboratory to investigate whether or not they could be used to guide missiles. Most important to brain itself are the exploits of Harvey Cushing, regarded as the father of modern neurosurgery, who served in the busy hospitals of war-torn France during World War I. Cushing brought home a stockpile of surgical techniques, adding to the shift of tides that was sending intellectual talent and spirit so aggressively to the shores of the United States. Just as Einstein was looking to the skies for answers about the universe, Cushing was pioneering the use of X-rays to look inwards at the very organ used in such cosmic contemplation—the brain.

The medical innovations that many of these people introduced were not without extensive experimentation and it was controversial operations like the prefrontal lobotomy that ushered in the risky confidence needed in approaching a procedure that would cut the corpus callosum. Presaged by animal experiments that traced the spread of epileptic seizures from one hemisphere to the other via the corpus callosum, William Van Wagenen spearheaded the first human split-brain operations in the 1940's. In what was sure to be a major disappointment, Van Wagenen did not see any signs of improvement in his patients. A fellow University of Rochester colleague named Andrew Akelaitis performed cognitive and behavioral studies on the patients only a few years later and found that the vast majority displayed no discernible differences as a result of the disconnection. The lack of any results put a halt to the callosotomy procedure and much of the enthusiasm about split-brain studies for over ten years.

Revival came through Roger Sperry at the California Institute of Technology in the early 1950's, sparked by his work that explored the transfer of information from the retina to the brain. His initial studies were performed on cats, and he found that by cutting the optic nerves that delivered visual information to the contralateral hemisphere—therefore restricting right-eye information to the right brain, and so too with the left—as well as cutting the corpus callosum, he was able to train each side of the brain independently. When one eye was covered, everything the cat learned was locked inside of the ipsilateral (same-side) brain hemisphere. This was precisely the double consciousness and curious coexistence by which the nineteenth century practitioners were so bemused (and for good reason!). There were now qualms about whether or not the previous procedures by Van Wagenen were sufficient to induce the desired effects, and more so, whether the studies from Akelaitis were truly representative of the patient's postoperative state, since Sperry's results were in such contrast.

The team at CalTech was ready for human trials with one major difference in their approach from past split-brain operations. Physiological studies were leading to the conclusion that cutting only the corpus callosum was not enough. A seizure often begins in only one hemisphere, and there is a certain threshold that must be exceeded in order for it to cross a midline barrier through structures like the corpus callosum. Once crossed, a feedback loop perpetuates the seizure into an uncontrollable monster of spastic activity. Sperry's team was planning an operation called a *commissurotomy*, which cut both the corpus callosum and the other major forebrain fiber tract called the *anterior commissure.*

The first patient who would undergo the procedure is known to the public only by his initials, "W. J." He was a World War II veteran who was injured by a variety of unfortunate events: first by a parachute jump gone wrong during a bombing raid in Holland, which rendered him unconscious and with a broken left leg, and later by the sharp blow of a rifle butt to the head while he was in an enemy prison camp. Stricken with unremitting blackouts and seizures for over twelve years after coming home— sometimes as many as ten per day—he was admitted to White Memorial Hospital in Los Angeles and handed over to the split-brain specialists. Previous diagnoses suggested that the problem emanated from the left hemisphere near Broca's Area, making it very likely that a localized procedure would result in the complete loss of speech. The commissurotomy was the only option that doctors had for saving W. J.'s speech while eradicating the seizures.

After several practice attempts on cadavers, doctors Joseph Bogen and Philip Vogel performed the surgery on W. J. on February 6, 1962. After making incisions into W. J.'s scalp to expose his skull, they sectioned off two separate pieces of cranial bone plates, exposing the underlying leathery sheath encasing the brain called the *dura mater* (literally meaning "hard mother"). After carefully making a U-shaped incision in his dura mater in both cranial cavities, they folded each piece of tissue to the side, which sufficiently exposed his brain. Maneuvering around the pericallosal arteries, the surgeons proceeded to cut the back half of the corpus callosum using a scalpel (the "blunt" method). They were more delicate in approaching the front half, making use of a microdissection instrument that applied a gentle suction to remove the brain matter. Last, the anterior commissure was divided, and using a mixture of sutures and staples, W. J. was put back together.

In the words of the surgeons, the operation "proved a boon" to W. J., as his seizures indefinitely disappeared. With no outwardly apparent deficits in W. J.'s personality, Sperry was quick to start his behavioral analysis, saying that the most remarkable effect of the procedure was in fact the "apparent lack of effect," and that what had been found in his animal studies had now proved true for man. Did that mean, too, that each hemisphere was now learning by itself, as in the cat studies?

Many more subjects who fit the criteria became candidates for the commissurotomy procedure, and those who went through with it were coming out with the same pleasing results as W. J. They would be known as the "California Series" of patients and participate in a number of tests by both Sperry and a graduate student named Michael Gazzaniga (who himself would play an instrumental role in furthering the understanding of the lone hemispheres).

Some of the tests were shockingly rudimentary. Imagine being blindfolded, with a paper lunch bag full of several random objects set before you—a toy plane, a rubber ball, a thimble, etc. You are asked to grab one, without looking, using your right hand. You feel its shape, its texture, and immediately recognize what it is.

Now, you must place the object back into the bag (still without looking). Finally, you are asked to find the same object, only this time using your left hand. A seemingly simple task, yet inconceivably difficult for a split-brain patient. The entire experience, for them, becomes encapsulated in the contralateral hemisphere from the hand that was performing the action, and afterwards, the memory is locked within the bounds of only one side of the brain. This is a failure in "cross retrieval." It's a loss in the fluid communication between the left and right brain that we unknowingly take for granted.

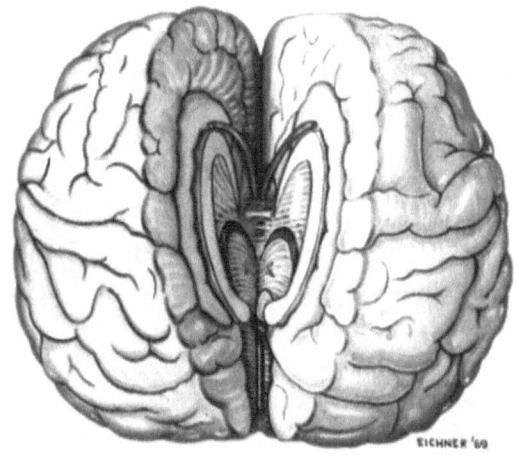

The Split Brain

Imagine the same experiment, only this time instead of being asked to name the object your right hand felt, you are simply asked to reach into the bag and grasp another object, now with your left hand. When the contents of the bag are thrown onto a table in front of you, and you are asked to find the object you picked, what happens? If you happened to be a split-brain patient, each hand would steer in separate directions—there may even be conflict between them as each side of your brain and your unforgiving subconscious realizes something is awry—yet, each hand eventually finds its *own* object. Surely floating at the borders of the sectioned cerebral commissures are signals that want to tell the other side about what it just held, but can't, and so a perplexing conflict arises.

Another blindfold experiment had a patient hold one hand straight in front of him, which was then posed by the researcher. The patient was asked to replicate the same hand shape with his other hand (called a "symmetrical hand pose" test). Once again, the inability to perform hemispheric cross-retrieval and integration denied the patient a proper sense of physical awareness and body poise, thus proving this to be an impossible feat for the patient. These are all signatures of what is called *disconnection syndrome*, onset by the commissurotomy. As Gazzaniga would go on to say, these were the moments in which "the modern split-brain story was born." There was a good reason to be excited, as this was the work that would later reward Sperry with a Nobel Prize, while cementing Gazzaniga as the foremost authority on split-brain research.

It was clear that memory and cognition had a distinct separation; the first signs of independent awareness, conscious thought, and strategy had been glimpsed when the two hands engaged each other in order to pick out their object from the table. Sperry was frank in his conclusion that the operation "left people with two separate minds, that is, two separate spheres of consciousness," indeed, just like the cats. One patient notably responded to an experiment with her left hand (controlled by her right brain) and immediately exclaimed, "Now I know it wasn't me who did that!"

How much merit did these statements have? This was precisely the jewel of a question that Justine Sergent researched before her infelicitous passing, picking up on the work of Sperry and Gazzaniga. It was quite obvious that consciousness—being aware of one's self—was apparently disconnected, but was everything? Since the commissurotomy left some of the subcortical commissures intact, it was conceivable that both hemispheres could still communicate, albeit only through the reptilian language of our brain's base structures—the rodent in our viscera.

Sergent deployed some ingenuity to the problem and developed a set of tests for split-brain patients to see whether or not they could integrate information that was distinctly held in the opposing hemispheres. She used an apparatus called a *tachistoscope* that would flash images separately on both the right and left sides of a screen, which were respectively projected to the left and right hemispheres as separate items. The novelty was that the left and right images were not sufficient in answering certain questions that Sergent would ask, and therefore had to be perceived and contemplated in tandem in order to provide an answer. For instance, she placed two short straight lines on each side of the screen, and in some instances, the two lines lined up with each other, and other times they were at random angles. The question asked to the split-brain subject was, "Are these lines straight?" Another test had an arrow appear in both the left and right visual fields, and the subject was asked, "Do these arrows meet head-on?" Other times, Sergent questioned the subject on whether the total number of objects on both sides of the screen was even or odd, or if the letters on both sides formed a word.

What Sergent found was quite astonishing. Split-brain subjects gave highly accurate answers to the questions, although when asked to consciously recall or articulate how they formed their answer, they were without proper explanation. There was something unconscious in nature that gave the subjects hints about whether to answer "yes" or "no" to Sergent's questions in a manner that was 75%-100% correct over an exhausting number of trials.

Yet another of her experiments displayed a photo of the Queen of England to only the right hemisphere, and then she asked the patient to verbalize (an action which relies on the left hemisphere) who was in the photo. The patient exclaimed, "I don't know," but said that whoever it was, "they don't have to work." Sergent went on to say that her results indicated the coexistence of *perceptual disunity* and *behavioral unity*, meaning that commissurotomized patients are more than just split down the midline. They are, in fact, subject and victim to a split of what we would call consciousness. Sperry and his team found similar results when they tried flashing a nude photograph to a woman's non-verbal hemisphere. She blushed but couldn't verbalize why she felt so flustered. Her "sneaky grin," as Sperry would recall, came from somewhere deep within which was quite obviously not a victim of the splitting in two.

We all experience this in some ways—through "gut" feelings—and popular books like Malcolm Gladwell's *Blink* are quick to expose the enigma within that provides subliminal hunches. The only difference is that you and I know what both sides of our brains are thinking, while the hunches that come into the minds of split-brain patients are no more than muffled, through-the-wall shouting of one hemisphere's information trying to get over to the other side.

* * *

Up until the mid-eighteen-hundreds, the inner workings of cells and the complex cascades of molecular reactions occurring within their membranes remained a riddle, wrapped in a mystery, inside of an enigma—to steal a phrase from Winston Churchill. The idea that all living organisms were driven by *élan vital*, the so called *vital force*, was employed as a means of acquitting God from scientific explanation, yet allowing a spiritual entity to still drive the doings of living organisms. Firmly pitted against the

nineteenth-century vitalists was the most prolific materialist up until his time, Hermann von Helmholtz, and along
with his brethren at the Berlin Physical Society, he signed
the following oath with his own blood!

> *[We pledge] to put in power this truth: No*
> *other forces than the common physical-chemical ones*
> *are active within the organism. In those cases which*
> *cannot at this time be explained by these forces,*
> *one must either find the specific ways or form of*
> *their action by means of the physical-mathematical*
> *method, or assume new forces, equal in dignity to*
> *the chemical-physical forces inherent in matter and*
> *reducible to the force of attraction and repulsion.*
> REYMOND-BRUCKE OATH

Helmholtz is really where the story of measuring
the mind begins, as psychology called upon physiology
to assist in mapping external-to-internal phenomena.
Although he was principally a physicist, he embodied the
category-spanning philosopher and practitioner that history deems so necessary. Anatomy had literally come alive
under his reign and had led many to question how exactly
the body communicated in and of itself.

The era of Helmholtz was pre-Morley-Michelson,
when energy was said to be encapsulated in atmospheric
aether, much before all of the connections between electricity, magnetism, light, and certainly the specifics of ion
channels and molecular chemistry were sorted out. Helmholtz's mentor Johannes Müller—the last of a dying breed
of vitalists—made very clear how the mysteries were perceived in the nervous system:

It is yet uncertain whether the nervous action,
which is propagated with such immeasurable
rapidity, is owing to the passage of an imponderable
matter along the nerves—whether the action of a
nerve separated from the nervous centers when
irritated, is owing to a current of such imponderable
matter taking place through it; or whether nervous
action consists merely in oscillations or vibrations
excited by the brain or the external stimulus, in
an imponderable nervous principle present in the
nerves; and it is a problem at present as little capable
of solution as the same question with reference to
light, namely, whether the theory of emanation or
that of undulation of light is correct.

We shall probably never attain the power of
measuring the velocity of nervous action; for we have
not the opportunity of comparing its propagation
through immense space, as we have in the case of light.
JOHANNES MÜLLER, ELEMENTS OF PHYSIOLOGY

Helmholtz was determined to disprove his mentor, Müller, and show that nerve conduction was neither imponderable nor immeasurable. Pulling from the famous experiments of Luigi Galvani, Helmholtz devised an ingenious rig to measure the transmission time of an impulse through the primary motor nerve of a frog's leg. Hooked into the nerve itself was a probe that was connected to a current source, which could stimulate the nerve on demand. The recording device he created was based on a galvanometer, which laid a needle on a revolving drum and recorded the current flowing through the probe and into the frog's leg (sometimes they show devices like these in movies when alien contact is made, or an earthquake is imminent). The true genius was that the position of the meter's needle was dependent on an electro-mechanical interaction. As the switch was depressed that connected the current through the circuit and into the nerve, the frog's leg was positioned in such a way that its contraction broke the same circuit as it flexed. Since the galvanometer was in series with the current flow, the deflection of its needle represented the precise moments that current was turned on by the human, and then off by the frog.

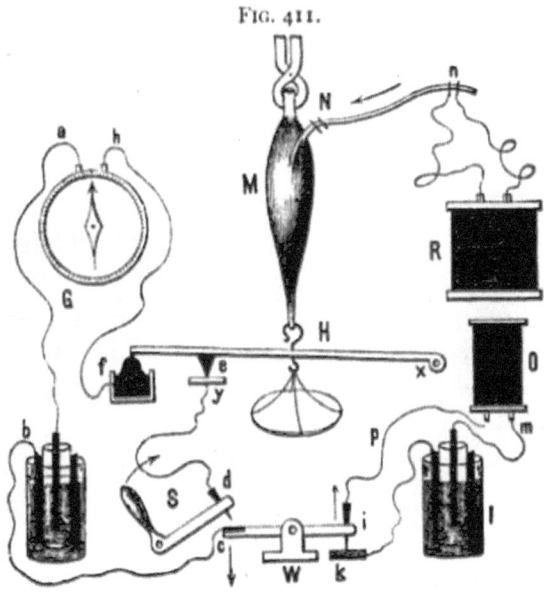

Helmholtz Method

A simple calculation involving the radius of the drum and the rate at which it turned decoded the markings into units of time, with a final resulting conduction speed of about thirty meters per second. In other words, a nerve the length of a football field would take about three whole seconds to transfer its signal from one end to the other.

Psychologist and historian Edwin Boring splendidly remarked on Helmholtz's discovery of nerve conduction, saying, "It brought the soul to time," as he measured an "essential agent of the mind in the toils of natural science."

It was the facts about timing that were most contended by the scientific community. Everyone had been sure that sensations were darn near immediate within the body! For a brief time, it caused a Humean stir, pitting what apparently *is* against what scientists had thought *ought to be*. This meant that every movement and sense in the body is a product of a signal which is considerably delayed, at least compared to Müller's notion that nerves spoke at rates up to, and perhaps beyond, the speed of light.

Eventually, the scientific establishment accepted Helmholtz's mental chronometry in full, and off went physiologists to determine reaction times of not just motor fibers, but the fibers that enlivened the mind.

The first of such cognitive reaction time tests asked subjects to identify a very simple stimulus, like a light bulb turning on, and then more difficult operations, like recognizing a shape in a drawing, or associating words. The results from these tests revealed what you already know: harder tasks take longer. Physiologists and biologists were feverishly decoding the nature of impulses in the brain to answer why this was. They found that the cell body (the neuron) was not the only thing influencing the vigor behind an impulse in the brain, and that the axons which carried the signal were just as important. However, it was found that the most significant timing variable in the brain's processing was due to synaptic junctions. It was, therefore, the extent to which one neuron is connected with another that was the basis for how long a process took to complete. More neurons means more synapses, and more synapses means more time.

With these discoveries of both mind and matter exposed, a rich materialism sprang to life that directly coupled the inherent delays of cognition with the complexity of neurons employed in the thought or action itself. It was now scientifically grounded that harder problems provoked more brain. This was a physical-mathematical model that broke no rules, worked without the assumption of any vital force, was completely carnal, and explained a rather elusive concept of the brain at that time.

A. T. Poffenberger incorporated this into a sort of systemology for interhemispheric conduction that is still used today. Established in 1912, *Poffenberger's Paradigm* leveraged the unique structure of the optic chiasm (the central meeting point of retinal signals) that relays visual information from the right half of each retina to the right half of the brain, and similarly with the left. Poffenberger found that if a stimulus is directed to the right side of the brain, the left side of the body has the capacity to react more quickly.

It takes a moment to understand how powerful this is. Imagine that you are sitting down in a completely dark room, facing forward, and are told to press a button with your left hand as soon as you see a light. In an instant—a matter of only a few milliseconds—a small flash occurs in your left peripheral vision. In that moment, you barely have time to saccade your eyes towards the light, and nowhere near enough time to move your body. The photons pass through both of your corneas and then refract on the lens sitting behind each pupil, making the leftwards light hit your rightmost retinas. The signal is then transmitted through your optic nerves into the optic chiasm where signals from similar areas of each eye are combined, thus beginning the process of creating one indistinguishable event for the mind.

After being paired together, the impulses race toward the right side of your brain, leaving your left side with no perception besides the empty and dark space that fills the right peripheral vision. The signals are then directed to a subcortical routing station called the thalamus, and from there, to the primary visual centers in the right occipital lobe. The occipital lobe is the most rearward of the four major lobes of the cerebral cortex; it makes decisions based on the information presented and then sends its riding orders to the motor cortices. Once your brain has decided what to do, your motor cortex springs to action, encoding an impulse targeted for the extremities. This burst of energy travels downwards and crosses to the other side of the body in a special part of the brainstem called the *pyramidal decussation* (literally meaning, to form an "X"), which is the reason we have contralateral control over our lower extremities. Finally, you press the button that your fingertips have been resting on.

Now imagine that you are asked to use your other hand with the same left-sided light flash. This time, your final button press is on the same side as the processing hemisphere, forcing your brain to talk amongst itself and invoke a mechanism that will send orders to the other side (if all this crossing seems confusing—don't worry, it certainly is). As expected, however, reaction times are longer in this second case because there are extra steps involved, which require and recruit more fibers and synapses in sending information across the brain's midline. In experimental terms, this is called the *interhemispheric relay time*, which usually adds 1-5 milliseconds to your overall reaction time.

Poffenberger's paradigm was the beginning of an empirical investigation into what may be running "under the hood" of us all. Using his timing scheme as a metric, it has been found that while split-brain subjects can trigger cross-hemispheric action, it takes considerably longer than in normal subjects.

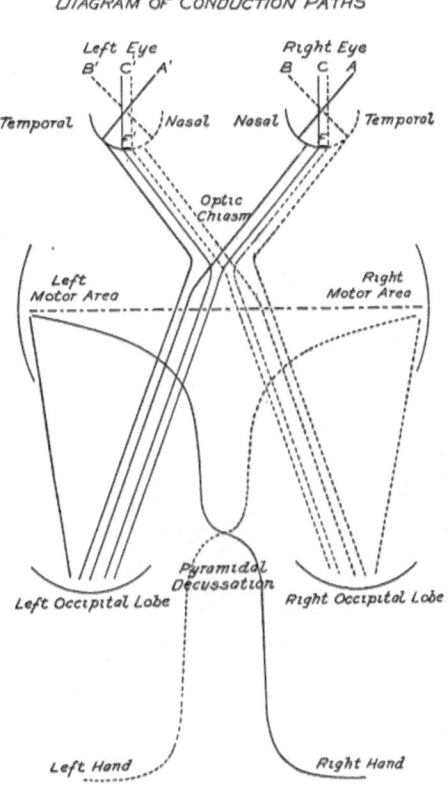

Conduction Paths in the Brain

This is evidence that there are, in fact, *extracallosal* pathways (besides the corpus callosum) that are integral in sending information from one hemisphere to the other. This type of tool allowed researchers like Sergent to hypothesize about what the brain is doing during the discrepant time. Is it yelling louder and louder for the other side to act? Is it strategizing in some way, within the depths of the sub-conscious, to relay information? Or is it simply bumbling around in mass hysteria and confusion?

The second major application of the paradigm is seen in subjects with callosal agenesis. While it also takes them longer to react, their reaction times are markedly shorter than the times represented by split-brain patients. This remarkable finding only corroborates some of the hypotheses about our oh-so-plastic brain, which is willing and able to cope with dysfunction, given enough time. If you are forced to be without your corpus callosum, it is best to get rid of it while you're young, so that your brain has time to creatively compensate.

A question deemed unexplainable was methodically approached by Helmholtz, and masterfully applied by Poffenberger. Certainly, these two were not the only ones who were involved in the story of turning the mysterious impulses which coursed through our fibers into under-standable phenomena, but they were monumental in many respects to science, and for our purposes, the study of lat-erality. Natural laws were beginning to be applied to the operation of the brain. The mind now had a substrate, a mathematical model, and there was an explanation—albeit rudimentary—for how signals move and at what rates. All that remained to be solved was what in the world the mind was doing during those few milliseconds that it took to react to a stimulus.

CHAPTER SIX

Split Function

There is as much difference between us and
ourselves as there is between us and others.
— MICHEL DE MONTAIGNE

In each of us there is another whom we do
not know.
— CARL JUNG

In general, we're least aware of what our minds
do best.
— MARVIN MINSKY

Stories at their greatest are religious...we are
story animals.
— YANN MARTEL

* * *

Famous for neurological investigations that, according
to him, were usually put into a drawer labeled "file and
forget," Dr. V. S. Ramachandran recorded a particularly
extraordinary result from his work with a commissurotomized patient. Using the same testing paradigms that were
used by Roger Sperry and Michael Gazzaniga in the 1960's,
Ramachandran isolated each hemisphere for a simple

question-and-answer session. First, he queried the right hemisphere (R.H.) in order to make sure it was willing and able to respond:

> Ramachandran: *Are you at CalTech?*
> RH Answer: *Yes.*
> Ramachandran: *Are you on the moon?*
> RH Answer: *No.*
> Ramachandran: *Are you in California?*
> RH Answer: *Yes.*
> Ramachandran: *Are you sleeping?*
> RH Answer: *No.*

Ramachandran then turned the topic to the divine in an attempt to probe the spiritual propensities of each hemisphere. With the option to respond with "I don't know" if the hemisphere ever felt on the fence regarding an issue, the conversation ensued:

> Ramachandran: *Do you believe in God?*
> RH Answer: *Yes.*
> Ramachandran: *Do you believe in God?*
> LH Answer: *No.*

The result was a person who was half-atheist and half-religious, half-sinner and half-saint, half empty and half full. If we were to maintain any corporeal expression of ourselves in Heaven and Hell, it begs the question of whether these souls are accompanied by only half bodies and whether the decussation in our brainstems would render the right half of the head and the left half of the body as those which are sacred. Our species as a population is critically divided on the subject of the divine, and more importantly, most individuals are guilty of having wrestled with belief in the supernatural or a higher order at one point in their lives. Can the conflict between opposing ideals and beliefs be related to our two brains, as Ramachandran's experiment would suggest?

As we will come to find, one of the major tenets of right hemisphere function is that it likes to tell stories. The only catch is that the tales it composes are only sometimes true and can often be elaborate embellishments of the real world—*just so* stories. Much like a movie inspired by a true story, the right brain works with bits and pieces of information, filling in the gaps when they are nonexistent, and sometimes when the real story is not sensational enough. It is a pattern generator creating lines of best fit. Our "sense-making machinery," as Nobel laureate Daniel Kahneman calls it, "makes us see the world as more tidy, simple, predictable, and coherent than it really is." This hypothesis is one we must consider, being that faith itself is defined as creating order out of discontinuities.

In a clear example of taking very small pieces and using them to make a whole—an exercise in pattern-making—we come back to W. J., the split-brain subject who was introduced in the previous chapter. He was given several cubes, each with two red sides, two white sides, and two half-red and half-white sides split diagonally. Next, he was presented with a series of printed cards which had different patterns that were all composites of the cubes he held in his hands. His task was to recreate the pattern shown on the card. The task was quite simply performed by W. J.'s left hand (controlled by the right brain). However, when asked to change hands, W. J. suddenly became perplexed and irritated by the situation. Every movement of his right hand came with marked hesitation. The video of his trials records a moment in which his left hand attempts to help out the right, only to be swiftly removed by the examiner.

One pathology that is almost entirely restricted to the right hemisphere—just as Broca's Aphasia is to the left—is called *neglect syndrome* (also sometimes called *hemispatial inattention* or *hemispheric neglect*). This is brought on by damage to the right parietal-occipital region of the brain, which is quite often the result of a tumor or stroke. Patients with neglect syndrome suffer attention-blindness to the left sides of their bodies, surroundings, and even to the spatial constituents within their memories that appear on the left side. When asked to copy a drawing from a sample, neglect patients will tend to draw the right half of the image correctly while completely neglecting the left side. I have been told of people with neglect who will only eat half of the food on their dinner plate until it rotated, after which they re-discover new food that has been in front of them all along.

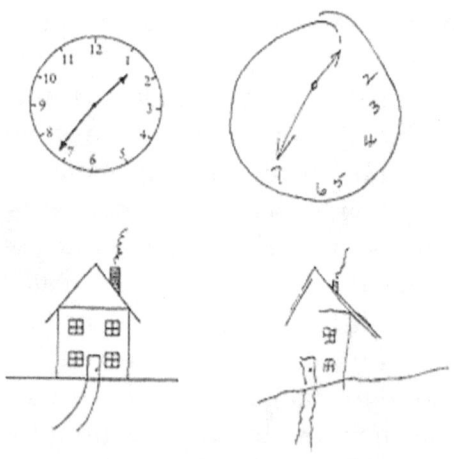

Neglect Syndrome Drawings

It has been shown that information about the left visual field does enter the brain and is processed to some extent, yet is trapped behind the walls of the subconscious by some mischievous mechanism. What this points to is an executive role of the right hemisphere in visualizing, patterning, and creating comprehendible models (a.k.a. stories) of the world. If this were not the case, neglect would be equally common in damage to the posterior regions of the left hemisphere, but it is not. With the right hemisphere intact, objects always seem to be whole and fully constructed, but when it is damaged, these objects are subject to a piecemeal representation.

I will bet that you can read the following sentence with full comprehension, and it is due to a phenomena called *closure*.

You acn flil in the gpas and erad bteween the leins.

Albeit a minor discomfort, you quickly realize that the middle details of each word are barely required. There is a part of us that finds it more efficient to just fill in the gaps, rather than to scrutinize the details. Apparently, closure in its most basic form is not exclusively a function of the right hemisphere. In 1972, one of Roger Sperry's associates, Colwyn Trevarthen, ran an experiment with split-brain patients that flashed half-pictures to each hemisphere. Despite the right hemisphere only being sent the image of half a bee, and the left hemisphere only half an image of a rose, each side of the brain interpreted the images as wholes.

The first time you catch a three-year-old coming out of the pantry with chocolate on their lips is a moment in which you realize storytelling is a complex phenomena that we are consistently working on. You might ask, "Were you eating cookies in there?" Immediately, the child is racked with confusion, as his eyes shift into a deer-in-the-headlights gaze, and then upwards and off to one side (we tend to look sideways when fabricating stories). As we grow into teenagers, we become much better at controlling our reactions and we lose the blank stare; we learn to control the momentary saccades that counterintelligence agents are trained to spot in liars. The stories get better, too, as we visually recall every place we have been in the past twenty-four hours and attempt to splice those visual scenes—removing the lewd ones, of course—into a cogent narrative to tell to parents.

When we thrust beyond the dealings of the real world and into pure imagination, much of the evidence points to the right side of our heads as our beloved regisseur: a producer who grips the storyboard of life and finds epic myths in the gaps of reality. This is a most important facet as to why the right hemisphere could be the harbor of the gods, those transcendent spirits that are omnipotent, unearthly, and in all objective terms, invisible. We have always been concerned with questions such as, "Where do we come from?" and "Where do we go?" Even now, our understanding of the expanding cosmos, of the sperm and egg, and of our beginning and ending as simply "star stuff" is not palatable as the full explanation of our existence. The story of a God, and the implied powers of His character, are a conceivable pattern to answer those unanswered questions. This explains why my dog sitting next to me never seems to pray to a shrine or sacrifice her toys to a doggy-deity; unlike humans, she has neither the fully-developed frontal cortex to meditate on such questions, nor the drastic asymmetries in her anatomical structure that play into their contemplation. My dog is seemingly not bothered by her fate because she has no concept of, or concern about it. Give any animal the notion of tomorrow, and I think religion is bound to appear on the scene. Modern agriculture guarantees us food for tomorrow, and our phratries assure us a mate, allowing us to devote nearly all of our time—if we so chose—to explore the mysteries of our before and after.

Scrutiny comes hand-in-hand with belief, especially in the brain, with the vital detail being in their order. The process of *believing* is made up of of three components: comprehension, acceptance, and unacceptance. René Descartes set up the philosophical scaffolding with these building blocks and suggested that we must first comprehend a precept or idea; only then can it be understood as something to be accepted or rejected. We must "know something truly before we can truly know it." In these terms, we all have the ability to decide what exactly gets imprinted on the mind, as if it were a precious piece of wax. The obvious problem is that comprehension takes time, and while this argument works for higher-level processing and contemplation to which humans have access, it conflicts with the primeval survival mechanisms that have been most important to the perpetuation of our species. The resource-intensiveness of the Cartesian process led Baruch Spinoza to make a revision in the sixteen-hundreds and propose that first we comprehend *and* accept; only after that do we unaccept, or reject, if necessary. According to his theory, everything flowing into our rational minds is immediately labeled "true," only to be pruned when there is either time or energy available. Spinoza's procedure is akin to a quality manager who has to oversee a conveyer belt of pastries running through a bakery warehouse. Their job is to let mostly everything through, and to spot and eliminate only the defects. Harvard psychologist Dan Gilbert reflects on what the Spinozian method suggests about human nature:

People are credulous creatures who find it very easy to believe and very difficult to doubt. In fact, believing is so easy, and perhaps so inevitable, that it may be more like involuntary comprehension than it is like rational assessment.

DAN GILBERT

If you have ever walked on a dirt trail or through a forest, I suspect that this simple scenario will allow you to evaluate these theories for yourself. You're walking down a path, the trees around you creating a canopy that dulls the sunlight. Suddenly, an abrupt shuffling of some branches and leaves behind you pierces your ears, and shortly after your perception. You snap backwards, your heart feels a jolt, and little known to you, your pupils have undergone a striking dilation. You survey the ground for a moment, and when the coast is clear, you turn back around and keep walking, as the hairs on the back of your neck settle down. It was probably a squirrel or gopher, maybe just a tree limb that fell; all of these possibilities are more likely than a mountain lion, so why did you react as if it *were* a mountain lion? Simply put, on the very off-chance that it was a giant cat, your foregoing of a quick reaction could cost you dearly—leaving you maimed or potentially a delectable dinner for some beast. The reason that we flinch and unconsciously enter survival mode—fight or flight—is due to the story-making machinery in our brains, concocting and predicting extreme situations, and sending them onwards as truths. The mental image of a giant spotted leopard pouncing on you, hungry for flesh, immediately prompts action. In the game of life, we don't lose points for flinching, but we do for inaction, and this is why our intuitive guesses must pass the barrier of the subconscious without proper scrutiny. Dr. Ramachandran explains how this happens on the streets of a big city too:

> *When a large threatening stimulus—say,*
> *an image of a menacing figure, perhaps a mugger*
> *looming toward me on the street in Boston—first*
> *comes to my brain, I haven't the slightest idea of*
> *what it is. Before I can determine, aha, perhaps*
> *that's a dangerous person, the visual information*
> *is evaluated by both the frontal lobes and the limbic*
> *system for relevance and sent on to a small portion*
> *of the parietal cortex, which, in conjunction with*
> *appropriate neural connections in the reticular*
> *formation, enables me to direct my attention to the*
> *looming figure.*
>
> V. S. Ramachandran, Phantoms in
> the Brain

The parietal region that warns of danger in the forest or on city streets is the same one that is damaged in neglect syndrome, the same one that is in charge of creating a foreboding mental state and fictitious images in attempts to fit a pattern. We are all subject to this biological automaticity, working presumably under a Spinozian mechanism, revealing the hidden nature of believing, which, as the evidence suggests, is housed in the right hemisphere. Sigmund Freud said that it will not be until we can escape the fear of death that we will not need this mechanism of "illusion." It is a cognitive system operating under automatic control, good at elucidating the surprising from the normal, inferring (and inventing) causes, believing rather than doubting, and tending to turn low probabilities into high ones. Our confidence in coherence is the basis of what we call *faith*, and there is mounting evidence that one side of our brain embraces it more fully than the other.

* * *

So far it should be clear why the right hemisphere ought to be, and presumably is, impressed by the idea of God. We might have some ammunition for answering Ramachandran's half-atheist experiment, but it hardly answers the question of what is going on between two brain hemispheres that might turn conflict into resolution. A person who maintains the interhemispheric commissures (like you and I) is prisoner to a constant debate being fought between the two sides of the brain, with the verdict emerging with apparent confidence. This obscure phenomenon of attention is explained by William James as "focalization," a concentrating and honing in on "one out of what seem several simultaneously possible objects or trains of thought." The reigning metaphor has always likened this to a spotlight of the mind, shining onto a great theater's stage, a product of the two hemispheres working in conjunction.

The left hemisphere is surely not fond of an invisible entity who is in charge; it likes the tangible world, and empirical data to be presented before it believes something. Oliver Sacks gives an account of a patient, Dr. P., who suffered from *visual agnosia*—the failure to make sense of the visual world—after injury to similar structures in the right hemisphere that produce neglect syndrome. Consequently, his left hemisphere had to take over, which thereby led to an overly objective view of the world. Sacks recounts the narrative of one session with Dr. P.:

> *I had stopped at a florist on my way to his apartment and bought myself an extravagant red rose for my buttonhole.*

Now I removed this and handed it to him. He took it like a botanist or morphologist given a specimen, not like a person given a flower.

'About six inches in length,' he commented. 'A convoluted red form with a linear green attachment.'

'Yes,' I said encouragingly, 'and what do you think it is, Dr. P.?'

'Not easy to say.' He seemed perplexed. 'It lacks the simple symmetry of the Platonic solids, although it may have a higher symmetry of its own. . . . I think this could be an inflorescence or flower.'

'Could be?' I queried.

'Could be,' he confirmed.

'Smell it,' I suggested, and he again looked somewhat puzzled, as if I had asked him to smell a higher symmetry. But he complied courteously, and took it to his nose. Now, suddenly, he came to life.

'Beautiful!' he exclaimed. 'An early rose. What a heavenly smell!' He started to hum 'Die Rose, die Lilie . . .' Reality, it seemed, might be conveyed by smell, not by sight.

I tried one final test. It was still a cold day, in early spring, and I had thrown my coat and gloves on the sofa.

'What is this?' I asked, holding up a glove.

'May I examine it?' he asked, and, taking it from me, he proceeded to examine it as he had examined the geometrical shapes.

'A continuous surface,' he announced at last, 'infolded on itself. It appears to have —' he hesitated —'five outpouchings, if this is the word.'

'Yes,' I said cautiously. 'You have given me a description. Now tell me what it is.'

'A container of some sort?'

'Yes,' I said, 'and what would it contain?'

'It would contain its contents!' said Dr. P., with a laugh.
OLIVER SACKS, THE MAN WHO MISTOOK HIS
WIFE FOR A HAT

This recount of Sacks' gives good indication that the left hemisphere is not only void of the basic visual skills possessed by the right, but that it struggles to combine pieces into a whole.

In one of the more famous experiments in laterality, the left and right hemisphere of a split-brain patient were queried about their desired career. The patient orally responded (from the left hemisphere) with "draftsman," and then used his left hand (controlled by the right hemisphere) to arrange a set of Scrabble letters that spelled "automobile race." What do you think this says about the two halves?

Michael Gazzaniga wanted to know how these different hemispheres—with their different personalities—could influence each other. He devised a stupendously simple projector that could flash cards with different images exclusively to each hemisphere of a split-brain subject. In one instance, the card presented to the left hemisphere was a chicken claw, and the one to the right hemisphere a snow scene. In front of the patient was a piece of paper with an array of images that were closely associated with the ones being flashed by the projector. The patient was then asked to make the correct association between the flashed images and the ones in front of him. It was a matter-of-fact process for the patient, as he reached forward and pointed to a chicken with his right hand, and a shovel with his left hand. The astonishing bit occurred when the researcher asked why he chose the items; his response was, "Oh, that's simple. The chicken claw goes with the chicken, and you need a shovel to clean out a chicken shed."

The left hemisphere was thus dubbed "the interpreter," and was happy to provide a narrative so as long as it was logical and in accordance with the actions, decisions, and beliefs of the right hemisphere. You will notice that in this case, the left hemisphere is the one telling stories, and some of the early work on split-brain research pinned the brain's ability to create order out of chaos in the left hemisphere. What appears to emerge over time, though, is a left hemisphere that has the ability to connect a linear narrative, but does not necessarily take leaps of faith. Another patient in the same double-image experiment was shown the word "music" to the left hemisphere, and "bell" to the right. As the patient's left hand (right brain) navigated to the image of the bell that sat in front of him, the left brain spoke up and conceded that the last music he had heard was from the library's bells just outside. The explanations given by each patient in these trials never indicated that the left hemisphere was capable of creating a fictional reality.

The left brain is where we find the associative center that relates our distinct bodies to the people, places, and things around us. This mirrors our earlier discussion of embodiment—knowing that our thoughts are contained in a body, and that the body is discrete. This is the mechanism that grounds our thoughts in reality and simulates what could actually happen in the real world. Chicken sheds really do get cleaned with shovels, and the man near the library really did have recollection of the bells ringing. These are examples of logical transferences across real categories of human experience.

The left brain is the philosopher and the right brain the theologian, both constantly churning away at hypotheses. The left is bound by a massive web of facts and forces perfect segues between each and every detail. The right can create highly-resonating stories, albeit without the requirement of evidence. Our one side is tugged on by rationalism, while the other ponders rectitude by means of faith. Sometimes the left is forced to concede because logic is simply not the right tool for the problem, and sometimes the right's fideism is brought to its knees by pure reason.

For insight into whether or not the "philosopher" and the "theologian" are ever to find equal footing, we turn to Ian McGilchrist, author of a most formidable book on the divided brain entitled *The Master and His Emissary*, who argues that the brain has been becoming more divided over time. Our intuition suggests that natural selection for a species as astutely wonderful as ours might be steered toward a certain elegance—maximizing symmetry, order, and oneness—but as McGilchrist suggests, and as our previous look into anatomy supports, this is not the case. McGilchrist proposes that our brain hemispheres are slowly becoming even more distinct characters in the theater of the mind, and that what emerges are two hemispheres that hardly recognize their actions as being unto the same body. He argues that this separation can explain the curious human condition with which we are concerned:

> *Phenomena that were previously uncomplicatedly experienced as part of a relatively unified consciousness now became alien. Intuitions, no longer acted unselfconsciously, no longer 'transparent,' no longer simply subsumed into action without the necessity of deliberation, became objects of consciousness, brought into the plane of attention, opaque, objectified. Where there had been previously no question of whether the workings of the mind were 'mine,' since the question would have no meaning—there being a cut off between the mind and the world around, no possibility of standing back from one's own thought processes to ascribe them to oneself or anyone or anything else—there was now a degree of detachment which enabled the question to arise, and led to the intuitive, less explicit, thought processes being objectified as voices (as they are in schizophrenia), viewed as coming from 'somewhere else.'*
>
> IAN MCGILCHRIST, THE MASTER AND HIS EMISSARY

When we think deeply and ponder the most essential questions, it could very well be that the answers from our right brain story-teller are perceived as coming from a higher order. After all, most religion followers could still be faithful without holy texts, relying purely on an indescribable certainty within them that can't be verbalized. The right hemisphere has long been discounted as the dunce of the two hemispheres because it is unable to express itself through formal language; however, we rarely give it reverence as perhaps the more important and righteous of the two sides.

This divine separation is certainly the point of view that Julian Jaynes takes in his classic, *The Origin of Consciousness in the Breakdown of the Bicameral Mind*. He argues that we are more symmetric than we may seem, in that there is an equally vibrant language center in the right hemisphere as well as the left. After all, if we evolved from symmetry, both hemispheres had similar capabilities at one point in time. Split-brain pioneer Eran Zaidel found that the right hemisphere probably has the language capability of a young teenager, and Jaynes thinks that this is a vestigial, whispering, distant voice that we turn into God. This voice has become so rejected by consciousness in favor of the left hemisphere's babble that we believe it to be outside of ourselves, making it a perfect candidate to fulfill the man-God relationship some people are so thirsty for. Research has found that stimulating the language area of the right brain evokes remarkable responses from patients. One saying that, "Again I hear voices, I sort of lost touch with reality," and another, "I could hear someone talking, murmuring or something."

There is a lot of evidence from case studies on epilepsy that corroborates Jaynes' position on the right hemisphere. When epilepsy attacks a particular part of the brain, it often removes that functionality altogether—much like a physical lesion to the brain structure. Patients with left-hemispheric temporal lobe epilepsy, affecting the primary language centers, appear to undergo a loss in left hemisphere function and subsequently, a *disinhibition* of the right hemisphere. This allows the right hemisphere to rise as the magistrate of the mind accompanied by a pathology known as *temporal lobe personality*. These people can find cosmic meaning in nearly anything (*Jesus Cheetos*, anyone?), and often have intense and ongoing episodes of spiritual enlightenment. Epileptic seizures can even lock in the once temporary changes onto the brain by means of *kindling*, which is a way that new pathways and connections are quickly forged out of irregular and overwhelmingly intense surges of activity. V. S. Ramachandran interviewed one of these patients who in response to, "Paul, do you believe in God?" responded with a puzzled look and said, "But what else *is* there?"

Schizophrenia is a condition that has been correlated to an increase in the responsibility of the right hemisphere, shown to take over some of the functions that usually reside in the left. The most notable symptom is that schizophrenics will hear voices, or see people and things which don't exist. However, another sign is that they begin to adopt a certain oneness with the world around them. One schizophrenic patient would say this about the erosion of their analog "I":

> *Gradually I can no longer distinguish how much of myself is in me, and how much is already in others. I am a conglomeration, a monstrosity, modeled anew each day.*
>
> ANONYMOUS PATIENT

This type of disembodiment can be tested in several ways which help doctors assess the mental state of a particular subject. One is known as the *Draw-A-Person Test*, which can be difficult for schizophrenics because they have lost their reference between themselves and others. They will forget to put bodies on people, neglect the arms, and make massive errors in judging proportion. One case of a schizophrenic patient recorded that they felt the world was pulling them apart, with their skin being the only thing holding them together, saying that "There is no connection between the different parts of my body." If you will remember, this is the same problem that children have during adolescence, which Piaget blames on an underdeveloped sense of self.

The famous ink blot tests that Hermann Rorschach developed are also a useful tool, as schizophrenics are often the recipients of *boundary loss*. They find that the edges of the image, which are inherently contrasting, appear to be fuzzy or simply nonexistent. What do you see?

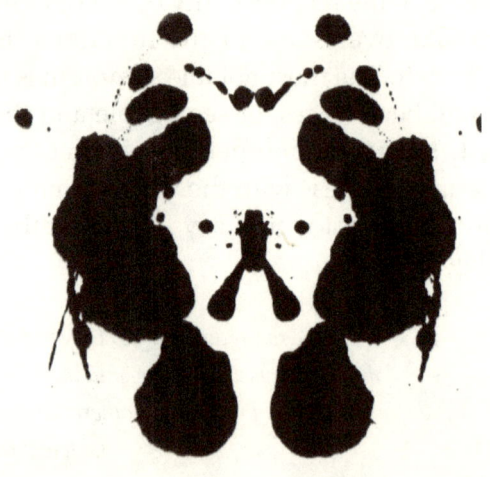

Rorschach Test

Neuroscientist Jill Bolte Taylor supplies us with a modern anecdote in her brilliant T.E.D. Talk, "My Stroke of Insight." She prefaces her story saying that the left brain is the conduit to reality: "It's that ongoing brain chatter which connects my internal world to my external world." Taylor woke up one morning to a slowly onsetting stroke in her left hemisphere, which soon took her verbal brain center offline. Her account of a reality dominated by the right hemisphere was that it was "euphoric," and propelled her into an expanding world without bounds between herself and the ground below her, the walls around her, and the objects near her. After spurts of regained consciousness, she recalls the moment when her right arm went numb, and she finally realized the peril of her situation, although she was oddly unaware of the urgency when immersed in the right brain's domain. She describes her journey to Nirvana with great emotion and detail, and speaks of the two cognitive minds that we all possess as together composing the life-force of the universe. "These are the *we* inside of *me*," she says.

Taylor makes it very hard to refute the theory of the right hemisphere as the home of our transcendent faculties. This is a materialist's viewpoint, for sure, but justifiable considering that Taylor regained normal consciousness following the removal of a golf-ball-sized blood clot from the right side of her brain. The good news is that this supports the idea that spiritual episodes or religious experiences are real, and perhaps there is more within us than we pretend to understand. However, it does not occlude the proposition that these flares of enlightenment could be transient ischemic attacks in the brain (otherwise known as *microstrokes*), bouts of schizophrenia, temporal lobe epilepsy, or any other of the numerous and sometimes fleeting ailments we humans are susceptible to. Even Plato recognized the connection between disease and divinity, calling insanity a "divine gift." He found no reason for the *t* that distinguished the greek word for prophetic, *mantike*, with the word for psychotic, *manike*.

There is a collective claim to be made in support of the divine voice that much of the world recognizes. Imagine you show up at a conference and are looking for your friend Bob. You walk around asking everyone in the room, "Have you seen my friend Bob?" Most people—in fact ninety-five out of the one hundred you ask—tell you with absolute certainty that Bob is somewhere in the room, and several of them mention that they just had a conversation with him. Now replace "Bob" with "God" and "conference" with "church." Belief can easily become a truth-by-numbers game. Without strong adherence to scientific principles, we would all be forced to search for "Bob's," and "Sam's," and "Billy's," depending on which congregation we found ourselves in. The true litmus test for an external God (or gods) can only be administered through rigorous comparative scrutiny. There are, however, hundreds of religions, with even more gods, which have all affected their followers in quite unique ways. In light of this we must wonder if these are all human ideations, and whether God has created man, or the other way around.

* * *

We have already been introduced to disconnection syndrome, which serves as an umbrella term for any pathology that is a result of a disturbance of the corpus callosum. As you can imagine, the list is as long as it is unsure. That is to say, depending on the affliction of the patient, he or she may exhibit many or none of the known symptoms. Sometimes these pathologies are easy to diagnose, as in the loss of language, hearing, touch, or through the incomprehension of words, or the inability to visualize both external objects (a plate on a table) and internal concepts (simple arithmetic). However, some are not so simple, yet emanate a wonderful glow onto the problem of the hemispheres.

Probably the most astonishing of these conditions is *alien hand syndrome*. This is like a special form of neglect syndrome that only affects the awareness of body parts. The patient is certainly aware of his or her hands moving about, yet they regard one of the hands as being alien and not a part of their body. They have both disowned the limb and forgotten about its history with them. Whatever is moving, grasping, and acting around them becomes something beyond their control, and beyond themselves as a whole. The patient is starkly aware of a limb, but speaks of it, and acts towards it, as if it were someone else's — a rather presumptuous *someone else* who is invading their personal space and wreaking havoc. Alien hands have been known to spontaneously choke people, including the sufferer themselves. On the lighter side, they have also acted as external safeguards, in one case throwing away a woman's cigarette before she could light it.

Sometimes limbs are *anarchic*. Instead of being fully neglected, the patient in these cases is fully aware of the limb being theirs, but the limb is unwieldy, uncontrollable, and in a sense, operates under its own government. For instance, the woman who had her cigarette thrown out referred to her hand as a "he," which suggests either that there had been a clear dissociation in the woman's mentality, or that she had a natural tendency to assign titles to her body parts. People with anarchic limbs are often not as confused as those with alien hands, at least not in the sense that there is a foreign hand flailing about, but are rather confounded that a part of their body has established goals without their conscious awareness of it. Imagine picking up the remote with your right hand and navigating to a TV channel; then, immediately afterward, your left hand picks up the remote and flips to another channel altogether. You pick up a book, and just as the right hand turns the page, the left hand tries to close the book entirely. One hand buttons your shirt; the other one unbuttons it. As soon as you get your pants pulled up with one arm, they are immediately pulled down by the other. One hand reaches through your shirt's arm hole, only to have the other hand promptly disrobe you altogether. The doors you open immediately get shut, and the chairs you pull out get pushed right back in. These are all real cases of the incessant annoyance that an anarchic limb can achieve. One can even draw a connection to the binary symptoms of obsessive compulsive disorder. Not surprisingly, it is usually the non-verbal hemisphere that is linked to the disruptive limb, which opens the possibility of an entire hemisphere gone rogue while unable to verbalize its intentions to the world.

The *horse race model* is a good place to start discussing how the two brain halves operate, and get to the bottom of the anarchic limb business. As soon as the brain is witness to some type of stimulus—according to this model—both hemispheres are off to the races, with the first to finish being the victor; this hemisphere gets to pride itself with handling the whole brain's response.

When I was in grade school, I remember we had a game that we would play during our math sessions. The teacher would ask us to clear our desks, and we were all handed a piece of scratch paper. Then, she proceeded to write three math problems on the board, and whoever finished solving them first would win some trinket from her top desk drawer. As she was writing the problems, the class hung on every new wisp and stroke of her hand, trying to make out the operand or the operator as quickly as possible, so the problem could start being solved. It was a race, but what we hardly knew was that it may not have been a race against the class, but a race against ourselves, from within. If you could look inside the brain, you might see two little people in each hemisphere, sitting at desks, with a math problem being fed onto the chalkboard of the mind.

Both of these little people race to get the answer, just as you yourself do in the real world. The first hand that is raised doesn't win a trinket, but gets to take control of the next steps, while the other one is shunned out of the sphere of consciousness.

While the left hemisphere may be implicated in math class for its logical abilities, most problems require skills of both sides of the brain. The classroom of life is inherently interdisciplinary, and the same problem can be solved in different ways. Albert Einstein relied on a *Gedankenexperiment*, or thought experiment, in formulating his hypotheses about the world. From his own writings, it is very clear that he was not the type of mathematician who discovered the secrets of our universe from rearranging equations, but rather thought about issues with his mind's eye. Our physical asymmetry and functional laterality become an asset when solving problems in life, and form the foundation of the horse race model. The hemisphere most capable will respond first and take control of the situation.

This model, though conceivable as a product of pure imagination or intuition, is found experimentally as well. It is called the *redundant-signal effect*. Studies show that if we are presented with the same stimulus—say a flash of light—to each hemisphere simultaneously, we respond faster than if the light were flashed to only one hemisphere. There is a sort of *redundancy gain* that occurs, which alludes to the apparent amplification of awareness and ability to respond that is accompanied by two stimuli rather than one.

Our brains are constantly working, and even in moments of utter deprivation, it is unreasonable to think that we will be able to reroute our entire inner circuitry to focusing on just a blinking light, or just a math problem. Each hemisphere's response time is not a static, known, unchanging number—it is dynamic, situational, with a sprinkle of randomness. This means that sometimes the right hemisphere will be quicker, and sometimes the left, depending on what other processes they are busy with. Some people would explain this by what is known as *finite resource theory*, which certainly has applications beyond biology. In terms of circuitry, there appears to be an "OR" gate controlling the response: for any number of inputs, as soon as one of them turns on (*this one* or *that one*), the output turns on.

The corpus callosum might be the track on which this horse race is run. When this structure is in good working order, everything works more smoothly, and a winner is picked unanimously. Without it, however, there is chaos, and it is quite difficult to distinguish a winner.

Researchers at the University of Michigan found something incredibly interesting about this model by studying split-brain patients. Previous studies (like those by Justine Sergent) show that basic information can traverse the brain's midline sub-cortically, via structures other than the corpus callosum. This was very much the case with W. J., who was able to respond to stimuli in both fields of view, but took about twice as long as a normal subject who had his or her corpus callosum intact. Despite the delay, Reuter-Lorenz discovered that W. J.'s redundancy gain was nearly eight times that of any control subject. This meant that W. J. was eight times *better* at solving a problem when both hemispheres were given the information, or inversely, eight times *worse* when only one hemisphere had the information. This is quite profound; it says that the modern-day corpus callosum enables us to better process in parallel. Without the corpus callosum, the brain has a difficult time coming up with answers, especially if only one side has information about the problem. In many situations, this means the corpus callosum is required to determine the winner of the horse race, and which side of the brain takes control of the response. You can imagine that without this type of referee, you would have the left and right brains pitted against each other, both rooting for their own unique answer. However, the mediation is not needed when the two sides are on the same team and in agreement, which, as Reuter-Lorenz shows, results in a quicker reply altogether.

This hypothesis also reintroduces the concept of symmetry to our discussion of the two brain halves. The parts of the brain below the corpus callosum appear to be more concerned about making the same decision before reacting. This reluctance is less apparent when the corpus callosum is fully connected, and, in fact, Reuter-Lorenz showed that there is a linear relationship between the physical amount of fibrous white matter in the corpus callosum and someone's ability to respond to asymmetrical (one-sided) stimuli. The more white matter we have, the quicker we respond. This finds agreement with nearly every proposition we have so far made about the human brain: it has been migrating ever more toward a love for asymmetry. Breaking symmetry has been a theme in our evolution for the past million years or so, since the time of our tool-making ancestor *Homo habilis*. The innards of our minds are conflicted by asymmetry, while the newer structures—the corpus callosum and our asymmetrical cortices—have found ways to transcend this confusion and deal with problems better and more quickly. Specialization is our claim to fame, and with it comes breaking old habits of mirrored processing between the two sides of our brain. If each hemisphere is to be allowed to process information independently, then it should follow that there would be a mechanism to break these old habits, and usher in the new—a brain able to think on each side, make decisions, and immediately act on them. This is the job of the corpus callosum!

In light of Poffenberger's findings from the last chapter, we should turn to the very curious fact that the time a brain takes to respond typically correlates to how much processing is taking place within it. How do we explain that the nuclei and networks of the brain below the corpus callosum become overly perplexed and devote a seemingly inordinate number of resources to asymmetric stimuli?

Imagine the environment in which our reptilian brain was nurtured. This was a development that occurred many millions of years ago when the terrain in which animals lived was either a vast sea, an open plain, or a dense jungle. All of these environments exhibited an expansive left-right symmetry. If you compared what the left eye saw to what the right one did, the images would be nearly identical, aligning themselves on the same horizon. Any break in this symmetry should indeed pique a creature's interest, likely being a barrier to movement, or worse, a predator. The older parts of our brains *should* be concerned about asymmetrical stimuli, because it represents something abnormal. Those animals had to put things of symmetry into the background—into the subconscious—far below any smidgen or curl of a cortex, or any substantial matter forming the corpus callosum. As the tree of life expanded from small mammals to primates, giving rise to our nearest ancestors, we began modifying our environment in such a way that asymmetry became more commonplace than symmetry. Not only were we using our two hands for different tasks, but we were building structures that had backs and fronts, doors that hinged, hearths that sat to one side of a water source, and beds that were particularly aligned. We began differentiating East from West, and started building rituals around cardinal directions. As our daily world became more asymmetrical, our brains had to make the momentous diversion from being bothered by asymmetry to embracing it. If our brains dedicated too much time to asymmetry in the world we live in today, we would be overburdened with nearly every daily action. If there is one explanation as to why we feel overjoyed and emotionally struck at beautiful places of symmetry such as the Parthenon, perhaps this is a calling from our ancient origins that enjoy one side being like the other.

There is some dance being had between the brain hemispheres, and at the center of it all is the corpus callosum. It can't be doubted that greater asymmetry requires greater stabilization, and it is reasonable to assume we have found the structure in charge of that. Our journey has taken us from the timelessly beloved and ardently posed question, "Can the mind be separated into two?" and replaced it with, "How do two minds make one?" If there is indeed an inter-hemispheric battle—to which the evidence appeals—we shall move forward and follow Carl Sagan's segue from the battlers themselves, to the arena in which they are brought together:

> *There is no way to tell whether patterns extracted by the right hemisphere are real or imagined without subjecting them to left-hemisphere scrutiny. On the other hand, mere critical thinking, without creative and intuitive insights, without the search for new patterns is sterile and doomed. To solve complex problems in changing circumstances requires the activity of both cerebral hemispheres: the path to the future lies in the corpus callosum.*
> CARL SAGAN, THE DRAGONS OF EDEN

Whether it is a dirty ruckus of horses trying to beat each other, or a symphony orchestra playing with perfect pitch and timbre, we know that some harmony must be in progress that enables cerebral rumination followed by pragmatic action. We are the only creatures that we know of so far able to create Feynman diagrams, contemplate the tragedy in Faust, and grasp the philosophy of Kant. Are these capabilities are the mellifluous amity of our brain halves, or simply, a resounding chatter.

*   *   *

Hitler's rise to Chancellor of Germany and his rallying of the Nazi Party came with the branding that we know so well today: the swastika. However, the symbol itself was not the Nazis' to begin with. Hinduism has held it sacred for thousands of years, it has been molded into pottery of the ancient African Kush Kingdom, embedded into Buddha's footprints at holy shrines, laid into the stones of the Romans, minted onto the silver of the Corinthian stater, and interpreted in writing long before it was clenched by corrupt supremacists of Nazi Germany. It was not until September of 1935 that the symbol took center stage on the national flag of Germany—painstakingly designed and dimensioned by none other than Adolf Hitler himself.

While horror ensued overseas, a young researcher in the United States was using the symbol in an experiment that would come to be one of the most well-known and highly-cited studies of all time. John Ridley Stroop had no idea that the swastika was going to become a symbol of tyranny and genocide; the thought of an American researcher upholding the symbol in the pursuit of science, while in Germany it was being displayed as the brand of an amoral nation, is either culturally, chronologically, or just coincidentally, amusing. This was quite surely one of the last studies in American psychology that would publish the use of the shape in its protocol.

Stroop was using the symbol in a series of studies that tested a person's ability to name the color of the text in which a word was written—the word itself being a color. You may have tried this before; read the following lists and say out loud the color of the text (gray or black), not the word itself.

BLACK

GRAY

GRAY

BLACK

GRAY

BLACK

BLACK

Stroop Effect

How did you do? If you are like any other human being, you had difficulty when the color (or shade, in this case) of the text was different from the color that the word described. This is the essence of the *Stroop Effect,* though it is normally presented with vibrant colors. For instance, the word "blue" would be printed in red, and the word "yellow" would be printed in orange.

There were two controls in the original experiment that provided a sort of baseline against which the research would be judged. The first were colors that were all printed in black ink, so the subject could read "purple," "green," and all the rest as he normally would. The second were swatches of plain colors that were used to get the subject familiar with simply naming the colors themselves. A small modification was made to these swatches to make them more representative of text—after all, the end goal was to see how the text itself conflicted with the color naming. Stroop changed the full-color swatches to swastikas, reasoning that their shape allowed color to be fully represented, while the edges and contrasts between the white background mimicked what occurs with real text.

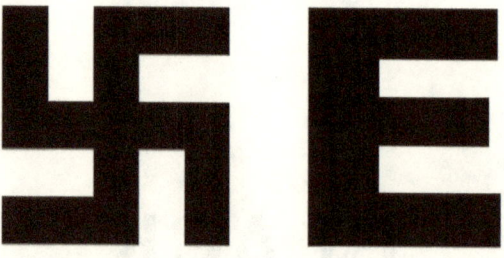

Swastika and the Letter "E"

The Stroop Effect became popular because everyone—including you!—is easily perplexed by the phenomenon. It challenges us to the core and questions how smart, aware, and capable we really are. While we hate to admit to being defeated by such a simple task, it is apparent that there is an instinctive force taking hold of our attention, and subverting it from the task we are asked to complete to something which is seemingly unrelated. Stroop was concerned with the obvious question: why does it take more time to name *colors* than to read *color names*? The answer comes only through a knock on the door of the unconscious; the automatic, and the unthinking selves within us.

The technical word for this effect is *interference*. Interference refers to the brain's predisposition to attend or succumb to an automated process rather than work on the problem contained in a higher level of consciousness. I tend to think of the Stroop test as a game show, with the very appropriate title *Name that Color!* in which contestants fail again and again at what seems to the viewers to be a mindless and easily won money-maker. What Stroop's interference stops short of answering is what processes are coming to life within our brain engine, and perhaps, between the two hemispheres of the brain.

When the test is administered to a subject's right visual field, which projects to the left hemisphere, the Stroop Effect (a.k.a. interference) is increased. In other words, the left hemisphere has a harder time naming the color than the right hemisphere. Why would this happen?

We have been tailored, taught, and trained since birth and through evolution to decode words when we see them. This automatic reaction is certainly the entity against which our brains are fighting when exposed to Stroop's test. We can probably all recall times in our lives during which we have been so vigorously trained to do something one way that we stumble when we have to do it differently. If you have ever followed up a mistake with the words, "I am just so used to doing it the other way," you know precisely what the left brain is going through when being asked to deal in colors rather than letters.

More support for this theory is offered by the two prominent Japanese writing systems, Kana and Kanji. Both of these are instantly recognized as foreign to a Westerner and rather indistinguishable, although Kana characters are simpler symbols made with two or three strokes of a pen, while Kanji writing is more intricate, like an Egyptian hieroglyph than anything else. The two systems represent the major difference between those writing systems that rely on graphemes, which combine characters into sounds to create words, and those that rely on logographs, which are symbols that in and of themselves represent a word or concept. English is clearly graphemic because we use letters to create words, and words to create stories, while the Egyptian hieroglyphs can contain an entire story within one or two symbols.

The Rosetta Stone was a major discovery in linking ancient logographs to modern graphemes. Found etched into the stone is the same story in three different types of writing—Egyptian hieroglyphs, Demotic script, and Greek—which allowed historians to finally bridge the gap between ancient and disparate cultures. As we will come to find, the Rosetta Stone is also a symbol of progressing from a language that speaks more clearly to the right brain, to one that is more clear to the left.

The Japanese writing systems prove valuable in split-brain explorations because of their internalized interpretations—Kana is primarily read and processed with the verbal left hemisphere, and Kanji is ideated and conceptualized with the visual right hemisphere. When studies tested the Stroop Effect against people who write in the Kana script, they exhibited the typical left hemisphere interference found in the original American studies. However, when people who were accustomed to Kanji were tested, the interference affected their right hemispheres. This means that the conflict within the brain was accentuated when the hemisphere that had to answer in terms of a color was also the one that had been trained to understand words. It was not just asking too much of the hemisphere, but it was asking for it to perform something uncommon. When the brain has been so routinely trained to decipher a word or logograph in a particular way, an abrupt detour from the norm occurs when something out of the ordinary is asked of it. This startles our automatism, and Stroop's paradigm is the classic example of how this can add milliseconds to our responses to a deceptively simple question.

A modified version of the Stroop task has given us some indication as to whether or not the brain is leveraging both hemispheres in the word-versus-color contemplation. Researchers placed a patch of color in one visual field and the word of a discordant color in the other. For instance, a small red patch of paper was placed to the left of the subject, and a word (printed in black ink) that spelled "blue" was placed to the right. The game was played the same way, with subjects being asked to name the color of the patch. The intention was to see whether or not the presence of the word, even when presented to an entirely different hemisphere from the color, would cause interference. Indeed, researchers found that it did, but only in the people with intact hemispheric connections. The Stroop Effect was just as strong for them even when the color and the word were separated into different hemispheres. This is the unconscious collaboration that is constantly churning within the connected brain.

*   *   *

The Japanese have a magnificent logograph that represents one's meaning in life—it translates to *ikigai*. This is said to be the thing that wakes us up in the morning and gives us the motivation to do our bidding for the day. In one study, 40,000 Japanese people, ages forty to eighty, were asked if they had an ikigai. After seven years, 95% of the people who responded "yes" were still alive and ticking, while only 83% of those who said "no" could claim the same. Apparently, having a purpose can save, or at least prolong, one's life. This is a familiar evolutionary credo and applies to life forms and their traits. The fibers that meet the two brain halves must be functionally significant, or else they would have slowly dissolved over time—but what do they truly live for? What is the ikigai of the corpus callosum?

生き甲斐

Ikigai Logograph

In neurological terms, the brain transfers information through either excitatory or inhibitory signals. Both are widespread and utterly important in the brain as mechanisms to shut off or turn on certain circuits or modules. The *patellar reflex*, commonly known as the knee-jerk reflex, is a ready example of these two signaling paradigms in the body. Being tapped on the patellar ligament (just below the knee) sends an impulse from the skin into a spinal cord segment in the lower back. The spinal cord itself contains many neurons, which form simple circuits and which work as fast-acting intermediaries between the senses and the brain, allowing us to react before actually thinking—these are "hard-wired", so to say. In this case, the connection at the end of the impulse is excitatory, which means that the pulse coming into the synapse (from the knee) will influence the neuron to generate its own impulse, thus propagating information forward. This spinal neuron is directly tied to the motor nerve of the quadricep, which travels out of the spinal cord down to the muscle itself and causes a muscle contraction, making the knee jump, twitch, or jerk. Pathways like these are known as *reflex arcs*. None of this would be possible, however, without an equally important inhibitory synapse to the motor nerves controlling the hamstring muscle. The quadricep and hamstring are antagonist muscles, which means that when one contracts, the other relaxes. You can imagine the stalemate to which our limbs might succumb if this were not a hard-wired feature of our peripheral nervous system.

The patellar circuitry is a relic of simpler times and a simpler architecture, when logic was much more dispersed throughout the body. For instance, an octopus has two-thirds of all its neurons located below the head, contained mostly in the spinal cord and arm segments. Ensembles of neurons within the periphery can account for why chickens are able to run around with their heads cut off, sometimes for days. Primates, like you and I, are the most head-centered of all animals, yet everything we do is still leveraging either excitation or inhibition between one cell and another.

Support of this view could be found in split-brain case studies themselves, which were authorized as a treatment for epilepsy, and aimed to stop a perpetual and amplifying feedback loop between each hemisphere. It makes perfect sense at first, until a greater appreciation for the function of inhibition comes into view. A very clear example is in the onset of shaking hands, known as tremor, as a result of Parkinson's disease. These tremors are thought to come not from neurons over-firing in the brain, as one might expect, but from inhibitory neurons under-firing. The circuits that would normally keep the tremors at bay are no longer online. This throws a stick into the road when thinking about what a commissurotomy truly does—is it stopping excitatory signals from flowing, or modifying inhibitory circuits that were misbehaving? Although the aesthetics of a single-ruled corpus callosum may be appealing, it is most certainly serving both excitatory and inhibitory roles.

The corpus callosum is in charge of the continuity between what each brain perceives, thinks, hears, and learns. We know that the connections are termed *homotopic*, meaning that similar areas from the left brain are connected to the right, but the precision of this relation is still coming to light. It could be that the mirror image of one hemisphere's activation pattern is mirrored to the other side, but only used when it needs help. This master-and-slave situation illuminates a potentially domineering relationship, which could swap back and forth between the two sides of the brain as a form of *metacontrol*. This type of control can also be scaled down, and would suggest that specialized functions are allowed to operate in one hemisphere, while a specific area of the brain is being suppressed on the other side. Both of these theories make up the macro and micro hypothesis of the most extreme types of hemispheric specialization.

Pondering the function of the corpus callosum as a whole is probably the wrong approach. In both of these cases, which have significant scientific foundations, the corpus callosum is implicated as the structure that irons out the inconsistencies of an asymmetric cortex. It does this by sharing information, restricting it, and mediating the responsibilities of cognitive function. This is why its physical structure has been correlated with everything from gender and handedness, to schizophrenia, depression, Alzheimer's disease, and dyslexia—it is much too complicated to have a single governing set of rules. It would be too much to say that the corpus callosum makes us human, but it certainly enables our condition. It is the most complex structure of fibers in our brain and continues to hint at our fate as both functional and dysfunctional primates.

* * *

I remember quite well how, when I was living in New York during the war as a refugee, I had dinner once with the great French composer, Darius Milhaud. I asked him, 'When did you realize that you were going to be a composer?' He explained to me that, when he was a child in bed slowly falling to sleep, he was listening and hearing a kind of music with no relationship whatsoever to the kind of music he knew; he discovered later that this was already his own music.
Claude Lévi-Strauss, Myth and Meaning

Sometimes, we hate what we love. There is a good chance that at one point in your life, your favorite song became utterly annoying, your dream job turned into a nightmare, or you began to loathe your best friend. But why does this happen? Certainly songs, jobs, and friends don't all change, so it must be you!

It turns out that some answers might be derived from a 1974 study conducted at Columbia University on professional and nonprofessional musicians. Researchers used an experimental paradigm called dichotic listening, which played distinct audio recordings to both the right and left ears. Just like vision and sensation, auditory processing is highly lateralized in the brain, most of it being sent to the contralateral side (the left ear is processed in the right hemisphere). Among a variety of people who had either very little, to a life's worth of experience with music, investigators found that these individuals' musical history dictated which hemisphere was better at interpreting musical sequences. While musically naive subjects deployed their right hemispheres onto the task, professionals used their left. The people with little experience were better at picking up melody and contours, whereas those with much experience had a knack for picking apart individual notes. The primary finding in these results comes as no surprise: the left brain likes pieces, and the right likes wholes.

More important, however, is the very essential observation about these subjects that the study makes: "as their capacity for musical analysis increases, the left hemisphere becomes increasingly involved in the processing of music." This means that when a person begins his or her journey into music, they use a different processing system than they will use later in their career. Professionals break down the music because they have to; they have been trained to be aware of the nuances, and even the errors. They become sticklers. They cannot see the majesty of the forest through the slightly-imperfect trees. An off-key note will ruin a performance for them just as badly as a misspelled word would ruin a novel for a grammaticist, or an awry brushstroke would ruin a painting for an artist. We would be fools to think this change only happens from right to left. This mode modification that we undergo is perhaps the reason that we find ourselves hating what we love. You might fall in love with art because you are gripped with emotion at the mixture of colors, the shadows, the contrast of lines, and the impressions of oils that you have at your disposal. And then, as you progress, you realize for yourself that those colors are tangible, those shadows are an illusion, those contrasts are just lines, those tones are arranged in a predefined way, and that the entire process of creating something beautiful has become robotic. This happens to many people—and in most cases it is because the romance of the pursuit is actually harnessed by the naiveté with which it is approached.

Long-term changes like this is neural plasticity hard at work. It is why new things become familiar. Without this mechanism our balance would be infantile, our speech continually slurred, and we could wave goodbye to any dexterity we wished we had. However, when it comes to love, we must find ways to make the familiar new. We are at war, comrades! Not with each other, but with ourselves.

CHAPTER SEVEN

Mapping

*Man is a plant which bears thoughts, just as a
rose-tree bears roses and an apple-tree bears apples.*
— ANTOINE FABRE D'OLIVET

*The modern geography of the brain has a
deliciously antiquated feel to it — rather like a
medieval map with the known world encircled by
terra incognita where monsters roam.*
— DAVID BAINBRIDGE

*Know then thyself, presume not God to scan.
The proper study of mankind is man.*
— ALEXANDER POPE

*　　*　　*

In 1865, the American archaeologist Ephraim Squier was
perusing antiques and artifacts in Cusco, Peru, surrounded
by some of the few remaining objects from the Incan
Empire. Among the faded pottery and dithered textiles, he
stumbled upon a most fascinating skull specimen. Through
the upper forehead was a two-by-two centimeter hole with
cross-hatched markings in the bone. Intuition told Squier
that there was a story here—something of cultural signifi-
cance—and so he cajoled the wealthy owner of the relics

into allowing him to take the skull back to the states for further investigation.

Squier himself wasn't able to find much on the topic of holes in the head. The only definitive text that addressed such matters was Samuel Morton's *Crania Americana*, yet any perforation or puncture to the head was simply attributed to warfare. Indeed, everything, from clubs and slingshots to swords and rifle butts, has been implicated in creating head injuries over the ages, but the breach in cranium that Squier was investigating had not resulted from any of these. Quite certainly, tools had been used on this head, and not weapons. After sharing the skull with several scholars in the United States, Squier remained without any convincing and cohesive theory that could explain this unique feature. It was then that Squier decided to ship the skull off to France, to be analyzed by none other than Paul Broca, who was now a world authority in medicine, anatomy, and anthropology.

Upon receiving the skull, Broca soon became convinced that the markings on the Peruvian man's skull were intentional and a result of what is called *trephination*. He came to a startling conclusion about the ancient Incas:

> ...*But here the trepan was performed on a point where there was no fracture, or probably even no wound, so that the surgical act was preceded by diagnosis....We are...authorized to conclude that there was in Peru, before the European epoch, an advanced surgery.*
>
> PAUL BROCA, A SINGLE CASE OF TREPANATION IN THE INCAS

No one had imagined that primitive tribes of the Andes could have been this competent in surgical matters. Adding to Squier's initial discovery, skulls were being unearthed all over France, revealing that European aboriginals had also made their way into the brain much earlier than anyone had previously thought. The real mystery remained: what *was* the diagnosis? After all, without knowledge of the vascular and ventricular systems, what had guided the crude tooling of these antediluvian surgeons?

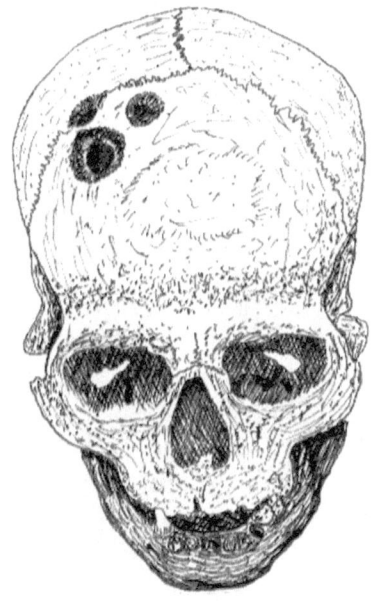

Trepanned Skull

Broca believed that the trepanning exhibited in these skulls was less about medicine, and more about ritual. Small polished and beveled cranial bones surrounded the gravesites of the trepanned skulls, often with extra holes in them that were thought to have been used for threading onto necklaces or bracelets. If a hole in the head was made to release evil spirits that terrorized a poor soul's body, the remaining bone fragments would have certainly been regarded as sacred amulets. The extracted bone would have been the plug to, as Broca phrased it, "a supernatural seal" in the head.

Western medicine has had trepanning protocols since the time of Hippocrates and Galen. These practitioners used trepanning for a variety of reasons—some still relevant today—including the removal of sub-cranial blood clots (a *hematoma*) and the repair of depressed bone fractures. From the frequency of skulls found in vastly separated regions of the world, one must conclude that the operation was being conducted for more reasons than just these few, and amazingly, with survival rates beyond what neurosurgery was able to achieve even into the nineteenth century. Close analysis of the other skeletal fragments and jaw remains reveal that it was not just men or women, young or old, who were having their heads carved open; it was quite nearly everyone.

Modern-day clues about what these cultures were thinking are slowly drifting into place, yet there is still some evidence of tribal medicine men prescribing trephination for everything from cancer to migraines. This ten-thousand-year-old practice casts an undeniable shadow onto our bodies by our heads, and suggests that we have always had the inkling to incline the status of our brains.

* * *

Without ever being told, do you think you would know that your head is your body's control center? Without ever seeing an image of a brain, or knowing what the spinal cord looks like, do you think you would still *feel* like everything coalesces up above? Would you be able to point to the area on yourself where the concept of *self*, and *I*, and *me* emerges?

As the evidence of trepanning alludes to, even without ample anatomical knowledge, we have for a very long time been thinking of our heads as the place where feelings, emotions, actions, and even gods and demons are interpreted or rendered. Trepanning is quite possibly the first indication of an attempt at localizing the functions of our nervous system. Galen had inadvertently stumbled upon some of these important functions while studying trepanned subjects:

> *Should the dissection be thus performed, then after you have laid open the brain and divested it of the dura mater, you canniest of all press down upon the brain on each one of its four ventricles, and observe what derangements have afflicted the animal...And when one pressed down upon that ventricle which is found in the part of the brain lying at the nape of the neck, then the animal falls into a very heavy and pronounced stupor...And if the incision should have been imposed upon the fourth... ventricle then the animal seldom returns to its natural condition.*
> CLAUDE GALEN, ON ANATOMICAL PROCEDURES

The very question of what happens when the brain is poked at and pressed about was best addressed many years after Galen by nineteenth-century French doctor and medical prodigy Pierre Flourens. He was a well-trained surgeon, as well as a prodigious investigator of brain function, coming into his professional career just as Gall's phrenology was undergoing large-scale skepticism, and just before

Broca entered the scene. He was pinned in history between these two theories of brain localization, yet he had his own reservations about whether or not the brain was as neatly mapped as some hoped it to be.

Flourens took an exploratory approach to disproving the whole idea of "'brain centers,'" and started slicing away at the brains of animals—from dogs to pigeons—all while recording their shifts in behavior. He stumbled upon the function of the vestibular labyrinth in the ear canal, which made all of his birds turn in circles when it was damaged, and he found that he could make respiration seize by ablating a specific part of the brain stem. Ironically, he had no concern about splitting the brain's function from the inside out, into areas of voluntary and involuntary control, but he stood firmly against a partitioning of the upper-most cerebral cortex. As he removed pieces of brain tissue during his investigation, he noticed no abrupt changes in the conduct of the animals, as one might expect if the brain was compartmentalized. Rather, what he found was a slow dithering of spirit in the animal, which decreased in proportion to the amount of brain matter removed—no matter where it was taken from. What emerged from Flourens was a doctrine of cerebral equipotentiality that was the antithesis to Gall's concept of localization. The cerebral cortex appeared to be a completely homogenous structure.

Eventually, the sciences behind these wide-ranging theories of localization were all forced to coalesce on middle ground, as the simplicity of Flourens' experiments were exposed, and the wild jumps beyond evidence from Gall were wrestled to reality. The best that could be said about functions such as language, perception, and executive control, is that they are *usually* controlled by a certain area of the brain. In order to know what was truly happening in each compartment of the cortex, humans would have to be the subjects, and probing would need to be done with needle-tip precision.

* * *

In the early nineteen-hundreds, electricity was fast becoming a force that neuroscientists needed to harness and command. Stimulation of the nervous system using electrical current was slowly growing into a viable method of mapping the brain to the body. The first mammalian experiments of this type was performed by Gustav Fritz and Julius Hitzig, during which the left brain hemisphere of a lightly anaesthetized dog was stimulated, resulting in the right leg producing a small twitch . As modest as that moment may seem, it was a turn in the tides—the scientist no longer had to permanently remove function from an animal or patient in order to study them. Electrical stimulation could both evoke brain cells and turn circuits "on," or create wide-scale disruption across brain regions and turn them "off."

Leading the charge on this innovative approach to the age-old question of brain localization was Montreal's first neurosurgeon, Wilder Penfield, working at McGill University in the 1940's. He had been investigating surgical interventions for epilepsy, which often included removing portions of the brain that were deemed hyperactive or otherwise corrupt. In some cases, the bad sections of brain were easily identified either by inordinate amounts of atrophy or excessive growth, but other times, the brain appeared completely normal. So how should he know what to remove?

Penfield began utilizing a small electrode to help guide him in surgery. With the patient under only mild anesthesia (or none at all), Penfield was able to interact with them during the entire process. The most obvious use for the electrode was to apply small shocks to areas of the motor cortex or language centers and survey the patient's response. In doing this, he could navigate around those important areas during his operation and avoid paralyzing his patient, or leaving them mute. The second, more inconspicuous, use of the electrode was to uncover what is called the *epileptogenic focus*, or the area of the brain that is most involved with the epileptic activity. Persons with epilepsy often have very vivid sensory or perceptual experiences right before the onset of their seizures; sometimes this is simply described as an *aura*, and sometimes it is a very specific memory, tune, or smell. Penfield found that if he could touch his probe to a place on the brain that also evoked that sensation or experience, he had identified the focus area, and knew exactly what tissue to remove.

In a famous Canadian television program, Penfield is depicted as standing over a patient with an electrode in one hand and the exposed brain below him. As he presses down the electrode to the brain surface, the patient speaks of the wonderful lights she is envisioning, and then, as the probe lands somewhere else, she remarks that she feels the sensation of cold water. Finally, as Penfield presses the electrode against a third location, her face succumbs to boding fear, then to excitement, and she exclaims, "Dr. Penfield, I can smell burnt toast!" This was one of the actual sensations that a patient of his would have experienced before an epileptic episode.

Although the actor in this skit became excited at the prospect of Penfield's knowing how to heal her and what brain tissue to remove, it was sometimes the case that the electrical stimulation actually drove the patient's brain further into its epileptic state, which was hardly ideal. This was not the first time that medicine has had to use humans as guinea pigs with undesirable outcomes. During one session, Penfield decided to use more electrical current than usual in probing the patient's left hemisphere (after all, no one knew what a "good" amount of current was yet). The patient first began to cry, and then to experience violent body spasms and convulsions on her right side. Her lips soon turned blue as froth spilled from her mouth, and she slipped away into a coma, dying three days later.

This technique, later known as the "Montreal Procedure," was only the tip of the iceberg for Penfield. With permission from both the hospital and his patients, he began taking extra time to survey patients' responses as he meticulously and minutely navigated different areas of the brain. He focused primarily on the motor cortex, since it was highly involved in epileptic activity anyway. This is a stretch of brain that lies mostly on the *coronal* axis, and would be covered from one ear to the other if you were wearing a headband. Prodding around the brain was a painstaking task not only for the surgeon, but also for the stenographer who was in charge of plotting all the data by hand.

What Penfield found after hundreds of trials was that there is indeed a map that runs along the vertical cortical crevice (once called the *fissure of Rolando*, now known as the *central sulcus*). The feet and legs have their control centers at the very top of the sulcus, and further down are motor areas devoted to the chest, the arm, the elbow, and the hand. At the very bottom of his map (lying just above the temples) are the neurons for the face, nose, and finally, the lips and tongue.

Although Penfield found some variation in the parts of the body that responded to specific areas of the brain, there was no doubt that he had overwhelming data in support of a highly localized cortex. From this, the *homunculus* (Latin for "little man"), one of the most famous figures in all of neuroanatomy, was formed. It depicts the relative size and position of brain nuclei as they run up the cortex. Electrical stimulation is still used today for locating important functional areas during surgery, although there is reason to believe we are entering an entirely new era of mapping and elucidating function from form.

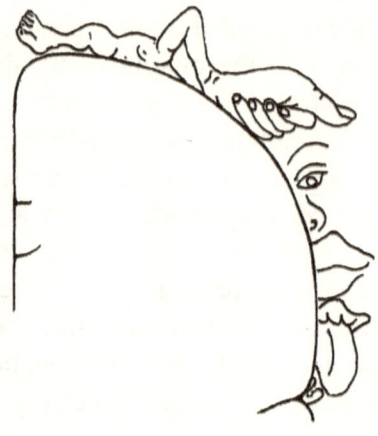

Homunculus

* * *

Former President George Bush Senior's proclamation that the 1990's was the "decade of the brain" was perhaps premature; however, the decade may have been an important turning point in the scientific community, the government, and the media, which all paved the way for what is happening today. Popular science magazines have seemingly converged on one icon for the epitomization of twenty-first century science and inquiry: the brain. Once a battle between the spinning electron and the smashed atom, the brain is now breaching a stronghold in public awareness, protected only by the mesmerizing spirale of the double helix.

Several ongoing projects around the world are supported by a sharp rise in the public's interest in both popular and clinical outcomes of brain research. Society has become equally enamored of the idea that we can learn more about love and creativity, as well as addiction and disorder, from pursuing the challenge of mapping the brain.

In 2005, the first modern mapping initiative took form in the Blue Brain Project, leveraging the processing power of IBM's supercomputers to simulate mammalian neural circuitry. By 2006, the research team had successfully simulated a chunk of cortex in the rat brain in charge of high-level sensory processing, called a *cortical column*, containing around ten thousand neurons and one hundred million connections (the human brain has about one million of these units). The novelty of this project lay in the complexity and caution taken, which gives each and every neuron a unique functional representation. Every single brain cell has its very own fingerprint—its very own distribution of ion channels that regulate charge, specific proteins and concentrations of chemicals, and a varying number of dendritic tree branches that gather signals from other cells. These distinctive dynamics control how often and how long the neuron would fire, which types of neurotransmitters are involved in the many operations, and other biological characteristics relating to both the internal and external structure. Everything from gene expression to large-scale connectivity is accounted for in one way or another, and one can see why this is not only a project pushing the limits of neuroscience, but also, computer power.

Beginning in 2009, the Human Connectome Project aimed to answer some of the more network-related questions about the entire brain's connectivity. Rather than focusing on individual neuron function, this five-year research project intends to discover macro-scale circuitry in the brain. After all, there is an inherent variability from my brain to yours which means that the number of brain cells, and certainly the connections they make, are similar, but not microscopically identical. From our interpretation of how the brain generates the mind, these connections are important on a very simple level: if two brain regions are connected, they likely collaborate in some way. Many different imaging techniques are being employed in mapping the brain, and if you haven't had one of these done on yourself, you probably will one day.

Functional Magnetic Resonance Imaging (fMRI) This measures the blood flow in the brain with a spatial resolution of about 1mm acquired over a period of about 1-2 seconds. Blood flow is presumably directly correlated with brain activity.

Diffusion Magnetic Resonance Imaging (dMRI) Also known as diffusion tensor imaging (DTI), this measures directionally-dependent variations in magnetic fields resulting from the layout of brain connections. This has been most notably used to reconstruct whole tracts of white matter across the entire brain.

Magnetoencephalography (MEG) This technique uses incredibly accurate sensors known as magnetometers to measure the naturally-occurring magnetic field fluctuations in the brain that result from neuronal processes and firing.

Electroencephalography (EEG) These measurements are typically taken by an array of sensors placed on some type of head cap for the purpose of mapping electrical impulses caused by an individual's brain activity.

Positron Emission Tomography (PET) This detects gamma rays emitted by a "tracer" that is introduced into a subject's bloodstream. PET scans are most commonly used to diagnose brain disease and dysfunction by analyzing blood flow and metabolic characteristics.

Between the Blue Brain Project and the Human Connectome Project, two of the major problems in neuroscience were addressed. If you think about the brain like a massive social network, the neurons are like each individual person. Everyone has a story, particular and personal attributes, and a unique way of dealing with situations as they arise. They also exist within a massive web of connections where a stream of minute-by-minute updates shape the global conversation and focus the so-called *collective unconscious*. The dynamics of such a system and interplay of individuals as they relate to the network as a whole is an incredibly challenging problem being undertaken by many researchers around the world. Uncovering the mysteries of these networks at the micro, meso, and macro levels requires a host of technologies and techniques.

Two additional projects along the same lines were launched in 2013, one by the European Union entitled the Human Brain Project, and another in the United States under the umbrella of the BRAIN Initiative (sometimes called the Brain Activity Map Project). Between these ventures, billions of dollars are expected to flow into the researching of brain systems in the next decade, with

President Barack Obama attempting to put some substance behind the statement made by George Bush Sr. more than twenty years ago.

The Switzerland-based, Human Brain Project mirrors some of the earlier computation objectives of other projects. Although it has many aims, one of the major ones is to be able to accurately simulate the effects of drugs on either a partial or a full-scale brain model. The difficulties that these teams will face cannot be underestimated. In R. Douglas Fields' book, *The Other Brain*, he reminds readers that the neurons and axons that generate and relay signals in the brain are not the only elements of this highly dynamic system. Supporting cells called *glia* make up a significant portion of the cells in the brain—perhaps up to 50%—and have been shown to play an incredibly vital role in brain function and dysfunction. Perfusing the brain with drugs has large-scale effects on the glial cells, altering first the external, and then the internal, environment of neurons. These are just some of the profound challenges that occur on varying scales for the Human Brain Project.

Quite different in purpose, the project in the States has modeled its initiative after the Human Genome Project, which notably came in under time and under budget. However, the ten-year timeline given for mapping the billions of neurons in the human brain is considerably lofty, and is likely going to require more funding before "mission accomplished" can be declared. It will start by picking up the systems and mapping research already done on much simpler organisms like worms and flies, and work its way into mammals, primates, and finally, human beings. One of the problems is that presidential initiatives will never rally support if their timelines are too lengthy, and their missions too small, so many have championed this as a great motion in support of science, albeit with some flare and pomp. This project's comparison with the Human Genome Project—despite its success—also clearly points to one of the issues with the public's interpretation of the intended goals. When the Genome Project began in 1987, many people imagined that decoding our genomes would be like unraveling a sacred scroll of secrets, or stumbling onto the Rosetta Stone of our biological makeup. However, what researchers later came to realize was that the story was much more complicated, and that bodily interactions outside of genomic processes have a big influence on behavior, health, and disease. The same predicament is likely to be faced by any project that sets out to map the brain. There are feedback loops in everything from molecules to behavior, and this is all shaped via learning and plasticity. These reactions, too, need to be modeled and accounted for before a "society of mind" can ever be realized. Luckily, included within the BRAIN Initiative are many proposals for nanotechnologies, noninvasive sensors, and biologically-scaled recording mechanisms which may be exactly what are needed to uncover the details of these mechanisms, which have up until now been hidden from view.

What this progress points to is not a decade of the brain, but more than likely, a century of the brain. The maps and models being produced will undoubtedly serve the scientific community in many ways that we are only starting to realize. However, the emergent properties of the brain will likely remain an enigma for some time. While physicists continue to work out how gravity interacts with the quantum world, neuroscientists will also continue their exploration of how molecules form the mind.

As a society, we are becoming ever more at risk of dying by means of our brains rather than our bodies, and the need is urgent for more robust tools to deal with this dilemma. Legs can be replaced with carbon fiber, arms with mechatronics, some organs with donors and some with artificial technology, but there is absolutely nothing that compares for the brain. If I had asked you two hundred years ago, "Would you rather lose half your leg, or half your memory?" you might have been conflicted, and even preferred losing your memory, because a cut-off leg was prone to infection and considerably more life-threatening. However, I dare say we have now entered an age in which we are less scared of our limbs succumbing to tragedy than our brains.

Understanding the network and functional topology of the brain will inevitably answer many of the major questions this book has explored. Brain-imaging studies are providing connection maps between brain regions with exponentially increasing resolutions. In the not-distant future, it is possible that we will all have access to three-dimensional scans of our own brains, and that machines will be portable and low-cost enough to download this data regularly, perhaps by visiting the local drug store, with each unit conveniently equipped with tools that provide automated checkups and diagnoses. Added to that, large open-source databases that are currently being established will provide crowd-sourced models of both healthy and afflicted patients, which will be free for you to use. The day is slowly approaching when some of the major advances in neuroscience may come from the equivalent of a Silicon Valley garage, utilizing off-the-shelf hardware and open-data repositories. There may very well be at-home kits for neuro-prognosis, just as there are now at-home kits that allow you to sequence your genome.

Mapping research will resolve particular questions that have long been asked about our asymmetry. We will have empirical data down to the smallest features of exactly how widespread brain asymmetry is, and what the effects are of either more or less asymmetry. The exact placement excitatory and inhibitory networks dispersed throughout the hemispheres have the potential to lay rest to some of the most puzzling neurological theories of the brains doubled structure and function. Does a healthy brain connect brain areas with more or less profundity than an unhealthy brain? Does stuttering really emerge from over-utilizing both brain hemispheres at the same time? Could the psychological conditions of a patient be related to a larger or smaller corpus callosum?

In 1984, Nobelist David Hubel remarked to his good friend, neurosurgeon Joseph Bogen, "The word Mind is obsolete." He was speaking in light of the growing amount of data coming from imaging and brain-mapping in the 1980's, assuming that the word "mind" would soon be expelled from the scientific lexicon. He said the very word mind "is like the word sky for astronomers." It was a warning that no matter how deep we dig, we may find ourselves even more stumped than before, rather than enlightened. What is *sky*, and where does it begin or end? There is no telling how much work is ahead, and whether or not these maps are the beginning or the end to a great revolution in neurobiology.

* * *

It was a Tuesday I would never forget. I met a handful of graduate students and postdocs at the Anatomical Science department on the University of Michigan campus. It was a short (and brisk) November walk around the corner from the biomedical research building where I was stationed. We had come to extract human brains.

The morgue was on the basement level of the building. Three of the walls were made of small bricks packed tightly together, laminated with a teal green that you only see on vintage kitchen appliances. The fourth wall was the body cooler to which detectives go directly when they visit the morgue on television shows. It was an array of fifteen square drawers, stacked three high, rising nearly to eye level, and spanning the entire width of the room. Everything that wasn't a brick or waste bin glimmered and shouted stainless steel; the tools were highly polished, the dissection table brushed.

The first-timers had been warned beforehand that if at any point there was a feeling of unease or queasiness, it was okay to step out of the room, and that there was no shame in that—as it "hits everyone at some point," I remember the professor in charge saying. What kept my nerves at ease was that these bodies and souls had known they were all coming here someday; "they were dying to be here," quipped a post-doc. The university's donation policy requires a doubly witnessed and willed consent before accepting a body, and not to be overly macabre, but I imagined that at some point, the head I was about to open had contemplated someone opening it, and had been okay with that. In spite of the irony, it is impossible to not feel a certain reverence for the individual and what remains of them.

We paired up into groups of two and were told that there were three buckets in the cooler marked "#12." We didn't have full bodies today, just heads. We opened the cooler door and rolled out the stretcher, which held the equivalent of three paint buckets, each with identifying labels on top and hazardous labels on the sides.

I worried about two things as we gripped the buckets and placed them on the floor. Firstly, that I was going to succumb to vulgar shock, quit pursuing my PhD, and never want to set foot near this place again. Secondly, I was even more worried about having the completely opposite reaction, that it wouldn't affect me at all. As we twisted off the tops of the buckets, I found that my emotions were pleasantly balanced in between these two extremes. The bucket held the head, from the neck upwards, with a small brass valve protruding from the carotid artery, which was used during the fixation process. As I picked up the head, it dripped some of the cold fluid that had kept it moist in the bucket. Slowly, I placed it onto the table, alongside two other heads of which other groups were in charge. I glanced around and moved with a hesitant deliberation, but every couple of minutes, the more experienced people in the room would say something funny, or pat me on the back. I was reassured that today was not only about harvesting brains for research, but an opportunity for teaching people like me about the internal anatomy of the human head.

Our subject was an old man, his hair buzzed and eyes closed, his wrinkles seeming to tell the same stories that an old pair of shoes does when you stare at them for long enough. Who was he? What did he do for a living? What was his family like? How did he die? These are all questions that are inevitable, but inconsequential to the operation at hand.

We had three tools in front of us: a scalpel, a saw, and a chisel. You realize very quickly upon entering any medical profession that rarely do you have the perfect tool; everything always seems "just good enough." It is hard to make perfect tools for imperfect and asymmetric bodies.

The procedure starts off a bit like a facelift, using the scalpel to pierce the skin on top and parallel to brow lines in the forehead; then, as if you were drawing with it, you cut a smooth curve upwards and around the ears, then down to the back of the neck, encircling the entire head with the same motion on both sides as to isolate a complete and continuous flap of scalp. It is worth taking a moment and tracing the route on your own head with your two fingers, as if you were brushing long hair off your face from your forehead to behind your ears. I slipped my fingers into the groove made by the blade on the forehead, and slowly peeled the flap of skin backwards, over the top of the skull. This is a feeling and sound that is unforgettable. Some tissue remains between the skin and bone, and as it separates, it is like pulling back the peel from an orange. My partner and I used some waxed string drawn in a circle around the cranium—at about forehead level—marking a straight path to cut by. We proceeded to take turns going around and around the skull with the saw until we began to puncture the brain cavity.

Once we were close enough, I took the chisel and pried off the top of the skull. Between the skull and brain sit several layers of protective tissue, the most prominent of which is the dura mater, the thin, leathery encasement that plays a vital role in keeping the brain locked in place and hydrated with fluids. With some finesse, the top of the skull pulled away, and as I set it down, it made me reflect on setting down the top of a coconut shell after halving it with a hacksaw. All I could think of was that I was reliving exactly what Andreas Vesalius had done nearly five hundred years before, and that our specimen was positioned in an eerily similar fashion to one of the famous depictions contained in his treatise, *De humani corporis fabrica libri septem*.

Confused and timid, I naively asked, "Now what?" The answer was precisely what I expected: "Grab it out of there!" And so our fingers reached around the frontal lobes as we pried upwards, making every attempt not to partially lobotomize our specimen. First, I held the brain while my partner cut the fibers and nerves connected to the cerebrum; then my partner held the brain while I reached way underneath the cerebellum and cut the brain stem. At last, the brain was free from its shackles, and with both of our hands on it, as if it were a precious stone, we raised it, feeling it free itself from the final bits of matter that had kept it suckered down.

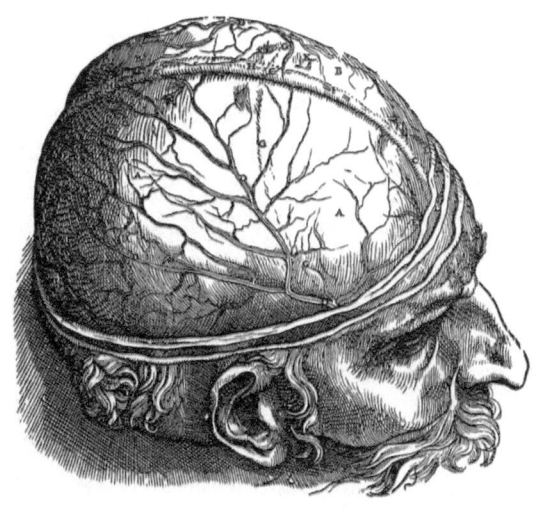

Drawing by Vesalius

There are certain things that are incredibly hard to describe, and this moment is certainly one of them. For anyone who has gone sky-diving, or traveled to a great museum, or hiked to a mountain's peak, you understand how words become completely inadequate when telling your story. What Schopenhauer says, that, "Thoughts die the moment they are embodied in words," becomes ever-so true. In my hands was the greatest mystery in the world, and the only organ that has ever named itself. It was the control center to our movements, actions, thoughts, and emotions. The beating of this man's heart ticked to the

197

rhythm autonomously created within this squishy mass. The books and films and art that he himself had witnessed were all once cataloged in here, and maybe they still are. Every hug he gave and every tear he shed was an expression of what I had in front of me. I was floored and overwhelmed; after all, there is one of these inside me, too.

I gently brushed my fingers down the left hemisphere's central sulcus, quickly moving past the motor cortex, and onto Broca's Area in the frontal lobe. Gently spreading apart the two hemispheres, I peered into the center as if it were a waterhole of infinite depth. My fingers ran along the middle edges of the cortex and onto the long span of brain called the cingulate gyrus. Finally, the tip of my index finger dove further, and was riding atop the corpus callosum. Without this bundle of white matter traveling across the midline, this book would not be here, and quite possibly, our affinity for dualism may never have come into existence. These are the millions of nerve fibers that keep right knowing from left, and right poised against wrong. They create split identities and binary harmonies.

After spending nearly five minutes rotating the brain about, inspecting every other crevice and groove, and tracing the veins and arteries as they sprouted and curtailed, I set the brain down. Everything that ever was and will be for that man—from cosmos to cognition—was just in the palm of my hands.

* * *

My anatomical escapade ends with slight irony and pity. We don't marvel at the motions of the heavens because we know the physics of them and have seen through satellite images precisely what other worlds look like. The effect of current through a wire on a compass is now taught in an afternoon of high-school science lab, but was the epitome of a revolutionary discovery just over one hundred years ago. Moving forward in time, those things which are not yet articulated—quantum particles, genomic interplay, and brain organization—fascinate our minds. Yet, one day, these breakthroughs in thought will become commonplace knowledge, simple facts of a discipline, and those of us enamored of our generations' mysteries will be labeled naive.

Take a moment to imagine a brain specimen transfixed on a solid countertop—still and silent. Slowly, anatomists, physicians, and philosophers from each generation, all the way back to the dawn of intellectualism, step forward onto the scene. Hippocrates looks over to Galen with a skewed expression of both awe and reversion. Galen glares at Descartes with suspicious intrigue, readying his nerves for exploring nature's blackest box. As Descartes surveys his peers, he looks at the brain, then upward, then back down, pondering where exactly the soul of this brain is at this place and time. Next, a modern anatomist approaches and doesn't waste time contemplating the taboo or lifelessness of the organ, but sees it as a network, once communicating and cascading in wondrous ways, as she narrow her eyes into the crevices as if she were seeking out individual neurons themselves. Finally, a scientist from the future approaches, completely and utterly unfazed. He sees in front of him an organ of which there are plenty. He knows how this brain can be repaired following any amount of damage. The thoughts and memories contained in this brain are backed up in massive data centers. Voices, movements, even individual nuances like ticks and flinches can be selectively reprogrammed into this vessel. The ways in which the molecular interactions translate into behavior, and how behavior is influenced by the environment, all have computational models and named laws for this future scientist. This very brain can be replicated — in part or in whole — given proper authority and the right medical facility. There is no repulsion, no inquiry, and no mystique anymore.

There are many scientific endeavors and collective research missions getting us ever closer to that future scientist. It does feel, however, that we may still be nearer to the Greeks than to having every detail of cognition formulated. We have a better understanding today than at any time in the past about how our brains function, and exactly what is happening between the two sides. Science will progress and discoveries will be made, yet there appears to be no progress on the front of knowing *why* we have two sides, and *how* exactly it alters our world. This book has surveyed the landscape of this conundrum — one that has been gnawing at the psyche for millennia. Within the pages of this book, we have been in search of the truths behind nature herself, the ones that almost certainly lie beyond science. The conflict between our hemispheres is the insanity that we are ignorant of on a daily basis, rendering us inherently subjective in our pursuits of objectivity, explaining why truth itself is so elusive.

We have subtly, yet enthusiastically danced around the topic of conflict, mediation, and middling extremes. It is a subject that is met most fiercely by religion, however there is much to be learned from our own anatomy. Finding a balance between opposing forces allows consciousness to emerge, but the struggle itself is necessary, it takes time, and is not always easy. There is yet to be a formula we can leverage in approaching such a problem, but there are hints we can take from our quick-acting subconscious, and the neural networks methodically pacing across our cortices. What we have learned is that resolve comes from a vigorous, and shall we say, scientific inspection of a problem. Early in this book we modified our approach to the hemispheres, from looking at both in conjunction, to just looking at one side in the famous split-brain experiments. Once it was clear *how* each hemisphere was different, there was ample ammunition to ask what the role of a mediator would, by necessity, be in many situations.

The jury is still out on whether our asymmetry is good for us, and if this war between both sides is at all healthy. Having two sides of the body and brain that are functionally unique appears to be an evolutionary advantage for humans; and functional specialization itself, at least on the surface, sounds like a good idea. In many ways, we have transcended the energy requirements of a battling brain through mechanized agriculture, and found artificial means of creating homeostatic environments through modern building materials and clothing. However, the evidence across the entire spectrum of natural elements shows that asymmetry, while useful, is almost always short-lived and extremely difficult to maintain. It appears that if we truly are on a one-way street to gross asymmetry, it may very well be a bumpy road.

I told you that while writing this book I asked many people about handedness, but there was a second question which I posed much more often, and with more interest. *What makes us special?* You and I—us humans. Sure, we are intelligent and creative, we have forethought, language, and imagination, which spawns economies, religion, and a concept of consciousness. We have indeed addressed many of these attributes throughout each chapter, some quite thoroughly, and there is only evidence that suggests we differ in degree from other animals, and not in whole. I had a priori decided when asking this question that I would argue whatever anybody said, because, I was certain that we just aren't that special. However, it occurred to me that the very topic of this book may be something that separates us from the beasts. We are the only ones who split ourselves, and the world around us into *two*. Our associations of right and left begin at birth, they arise in all environments and geographies, and we pass the tradition down extrasomatically through generations. Either our biology has modified us, or we have modified our biology; in either case, no other creature shows such bias. We are prone to losing our most valued faculties upon injury to a very specific side of our body — speech, comprehension, spatial awareness, and dexterity can selectively be abolished. No other organism is so fragile, but then again, nothing else is quite human.

Our brains can be divided in many different ways, but for some reason dipoles and dualisms are always most useful. Scientists like to think about how the conscious mind is elevated from its subconscious underpinnings. The faithful are committed to an earthly component and a transcendental one. Of course, most familiar to us all is the split down the middle, where unique identities are forged for each side, that we call *left* and *right*. Which one of these divisions is most important? Are there functional trinities or quadralities that are equally valid? These are questions I leave to you. They are the unknowns that keep our hearts beating and our souls occupied in lust for *more*.

Bibliography

PREFACE

Sherrington, Sir Charles. "Man on His Nature, 1940." UniversityPress, Cambridge (1963).

CHAPTER 1

Alan Bates. The Anatomy of Robert Knox: Murder, Mad Science and Medical Regulation in Nineteenth-Century Edinburgh. Sussex Academic Press, 2010.

Bergman, Jerry. "Darwin's Teaching of Women's Inferiority." Darwin's Teaching of Women's Inferiority. Web. 30 Mar. 2014.

Berker, Ennis Ata, Ata Husnu Berker, and Aaron Smith. "Translation of Broca's 1865 report: localization of speech in the third left frontal convolution." Archives of Neurology 43.10 (1986): 1065.

Bischoff, Christian Heinrich Ernst. Some Account of Dr. Gall's New Theory of Physiognomy: Founded Upon the Anatomy and Physiology of the Brain, and the Form of the Skull. Treuttel, Würtz and Richter, 1828.

Calvert, George Henry, ed. Illustrations of phrenology. W. and J. Neal, 1832.

Combe, George. A system of phrenology. William H. Colyer, 1843.

Dronkers, Nina F., et al. "Paul Broca's historic cases: high resolution MR imaging of the brains of Leborgne and Lelong." Brain 130.5 (2007): 1432-1441.

Franz, Shepherd Ivory. "New phrenology." Science 35.896 (1912): 321-328.

Gall, F. J. "The Influence of the Brain on the Form of the Head." (1835).

Gallardo, Susana. "Gender, Science & Objectivity." Stanford University, n.d. Web. 30 Mar. 2014.

Gould, Stephen Jay. The mismeasure of man. WW Norton & Company, 1996.

Mac Gregor, George. The History of Burke and Hare and of the Resurrectionist Times. Library of Alexandria, 1884.

Phrenological Society (Edinburgh, and Scotland). Transactions of the Phrenological Society. 1824.

Rosner, Lisa. "The Anatomy Murders." (2009).

Roughead, William, ed. Burke and Hare. W. Hodge, 1921.

Sagan, Carl. Broca's brain: Reflections on the romance of science. Random House LLC, 1980.

Schiller, Francis. Paul Broca: Founder of French anthropology, explorer of the brain. Oxford University Press, 1992.

Shaffer, Michael, and Michael L. Veber, eds. What Place for the a Priori?. Open Court, 2013.

The Phrenological Journal, and Magazine of Moral Science, for the Year 1839. 1839.

The Phrenological Journal, and Magazine of Moral Science, for the Year 1840. 1840.

CHAPTER 2

"Meet Washoe." Friends of Washoe. Web. 30 Mar. 2014.

Blumberg, Neil. "Blood work: A tale of medicine and murder in the scientific revolution." The Journal of clinical investigation 121.6 (2011): 2063.

Butler-Bowdon, Tom. Fifty Psychology Classics. Nicholas Brealey Publishing, 2007.

Chance, Steven A., and Timothy J. Crow. "Distinctively human: cerebral lateralisation and language in Homo sapiens." J Anthropol Sci 85 (2007): 83-100.

Corballis, Michael C., and Ivan L. Beale. The ambivalent mind: The neuropsychology of left and right. Chicago: Nelson-Hall, 1983.

Coyne, Jerry A. Why evolution is true. Penguin, 2009.

Darwin, Charles. The descent of man. Digireads. com Publishing, 2004.

Darwin, Charles. The origin of species. John Murry, 1929.

Davidson, Richard J., and Kenneth Hugdahl. Brain Asymmetry. Cambridge, MA: MIT, 1996.

Fouts, Roger S., and Deborah H. Fouts. "Chimpanzees' use of sign language." (1993).

Geschwind, Norman. Language and the brain. WH Freeman, 1972.

Greenspan, Stanley I., and Stuart Shanker. The first idea: How symbols, language, and intelligence evolved from our primate ancestors to modern humans. Da Capo Press, 2009.

Holloway, Ralph L. "Human paleontological evidence relevant to language behavior." Human neurobiology 2.3 (1983): 105-114.

Holloway, Ralph L., and Marie Christine De La Costelareymondie. "Brain endocast asymmetry in pongids and hominids: some preliminary findings on the paleontology of cerebral dominance." American Journal of Physical Anthropology 58.1 (1982): 101-110.

Huxley, Thomas H. Man's place in nature. Courier Dover Publications, 2013.

Jablonka, Eva, and Marion J. Lamb. Evolution in four dimensions: Genetic, epigenetic, behavioral, and symbolic variation in the history of life. MIT press, 2005.

Joseph B. Hellige. Hemispheric asymmetry: What's right and what's left. Vol. 6. Harvard University Press, 1993.

Kurzweil, Ray. How to create a mind: The secret of human thought revealed. Penguin, 2012.

Leakey, Louis SB, Phillip V. Tobias, and John R. Napier. "A new species of the genus Homo from Olduvai Gorge." Nature 202.4927 (1964): 7-9.

Lindly, John. "Archaeological Perspectives on the Origins of Modern Humans: A View from the Levant." American Anthropologist 103.1 (2001): 224-225.

Oakley, Kenneth Page. "Man the tool-maker." (1957).

Pinker, Steven. The blank slate. Southern Utah University, 2005.

Roche, Hélène, et al. "Early hominid stone tool production and technical skill 2.34 Myr ago in West Turkana, Kenya." Nature 399.6731 (1999): 57-60.

Sagan, Carl, and Ann Druyan. Shadows of forgotten ancestors. Random House LLC, 2011.

Sagan, Carl. Dragons of Eden: Speculations on the evolution of human intelligence. Random House LLC, 2012.

Smaers, J. B., and C. Soligo. "Brain reorganization, not relative brain size, primarily characterizes anthropoid brain evolution." Proceedings of the Royal Society B: Biological Sciences 280.1759 (2013).

Thompson, Darcy Wentworth. "On growth and form." On growth and form. (1942).

Tobias, Phillip V. "The brain of Homo habilis: A new level of organization in cerebral evolution." Journal of Human Evolution 16.7 (1987): 741-761.

Tobias, Phillip V. "The species Homo habilis: example of a premature discovery." Annales Zool. Fennici. Vol. 28. 1992.

Toga, Arthur W., and Paul M. Thompson. "Mapping brain asymmetry." Nature Reviews Neuroscience 4.1 (2003): 37-48.

Wilson, Darrell M., et al. "Growth and intellectual development." Pediatrics 78.4 (1986): 646-650.

CHAPTER 3

Baumeister, R. F. (2008). Free will in scientific psychology. Perspectives on Psychological Science, 3, 14-19.

Close, Frank. Lucifer's legacy: The meaning of asymmetry. Courier Dover Publications, 2014.

Criminal man. Duke University Press, 2006.

Feynman, R. P., R. B. Leighton, and M. Sands. "The Feynman Lectures." The Feynman Lectures 2 (1964).

Francis Crick. Astonishing hypothesis: The scientific search for the soul. Simon and Schuster, 1995.

Gardner, Martin. The new ambidextrous universe: Symmetry and asymmetry from mirror reflections to superstrings. Courier Dover Publications, 2005.

Marler, Peter R. "The Marvels of Animal Behavior." Natural Science Library. Natl. Geog. Book Service, Washington, DC (1972).

McManus, Chris. Right hand, left hand: The origins of asymmetry in brains, bodies, atoms and cultures. Harvard University Press, 2004.

Møller, Anders Pape, and John P. Swaddle. Asymmetry, developmental stability and evolution. Oxford University Press, 1997.

Møller, Anders Pape, and Mats Eriksson. "Pollinator preference for symmetrical flowers and sexual selection in plants." Oikos (1995): 15-22.

Wright, Robert. The moral animal: Why we are, the way we are: The new science of evolutionary psychology. Random House LLC, 2010.

CHAPTER 4

"Ardhanarishvarastotram with Meaning." Sanskrit Documents Collection. N.p., n.d. Web. 30 Mar. 2014.

"Buddhism - The Middle Path ." BuddhaNet, n.d. Web. 29 Mar. 2014

Aristotle, Metaphysics. "trans. Hugh Tredennick." Aristotle in Twenty-three Volumes [1933; Cambridge: Harvard University Press, 1989], V 29.4 (1933): 287-89.

Contemporary Comparative Side-by-side Bible: New International Version, New King James Version, New Living Translation, The Message. Zondervan, 2011.

Coren, Stanley. The left-hander syndrome: The causes and consequences of left-handedness. Simon and Schuster, 2012.

Dennis, Wayne. "Early graphic evidence of dextrality in man." Perceptual and motor skills 8.h (1958): 147-149.

Eagleman, David. "Incognito: The secret lives ofthe brain." Toronto, ON: Viking (2011).

Edwards, Betty. Drawing on the Right Side of the Brain. ACM, 1997.

Evans-Pritchard, Edward Evan. Right & left: essays on dual symbolic classification. Ed. Rodney Needham. Chicago: University of Chicago Press, 1973.

Farrell, William S. "Coding left and right." Journal of Experimental Psychology: Human Perception and Performance 5.1 (1979): 42.

Harrington, Anne. Medicine, mind, and the double brain. Princeton: Princeton University Press, 1987.

Hertz, Robert. "The pre-eminence of the right hand: A study in religious polarity." HAU: Journal of Ethnographic Theory 3.2 (2013): 335-57.

Hunt, Morton. The story of psychology. Random House LLC, 2009.

Iaccino, James F. Left brain–right brain differences: Inquiries, evidence, and new approaches. Lawrence Erlbaum Associates, Inc, 1993.

Kennedy, J. B. "Plato's Forms, Pythagorean Mathematics, and Stichometry." Apeiron 43.1 (2010): 1-32.

Laponce, Jean A. Left- and Right-Handedness Study, 1970.

Lévi-Strauss, Claude. Myth and meaning. Routledge, 2013.

Liang, Henghao. "Jung and Chinese Religions: Buddhism and Taoism." Pastoral Psychology 61.5-6 (2012): 747-758.

McManus, I. C. "Handedness in twins: a critical review." Neuropsychologia 18.3 (1980): 347-355.

Oldfield, Richard C. "The assessment and analysis of handedness: the Edinburgh inventory." Neuropsychologia 9.1 (1971): 97-113.

Peek, Fran. The real rain man, Kim Peek. Ed. Stevens W. Anderson. Harkness Pub. Consultants, 1996.

Piaget, Jean, and Margaret Trans Cook. "The origins of intelligence in children." (1952).

Previc, Fred H. The dopaminergic mind in human evolution and history. Cambridge,, UK: Cambridge University Press, 2009.

Schiefenhövel, Wulf. "Biased semantics for right and left in 50 Indo-European and non-Indo-European languages." Annals of the New York Academy of Sciences 1288.1 (2013): 135-152.

Springer, Sally P., and Georg Deutsch. "Left brain, right brain: Perspectives from cognitive neuroscience." (1998).

The Sutra of Hui-neng, grand master of Zen: with Hui-neng's commentary on the Diamond Sutra. Shambhala Publications, 1998.

Wain, Harry. "The story behind the word." Springfield, IL: Charles C. Thomas (1958).

CHAPTER 5

Bear, Mark F., Barry W. Connors, and Michael A. Paradiso, eds. Neuroscience. Vol. 2. Lippincott Williams & Wilkins, 2007.

Bogen, Joseph E., and PHILIP J. Vogel. "Cerebral commissurotomy in man." Bulletin of the Los Angeles Neurological Society 27.4 (1962).

Damasio, Antonio R. "Descartes' error: Emotion, rationality and the human brain." New York: Putnam 352 (1994).

Fox, Peter T. "Why Did a Brilliant Researcher Choose to Take Her Own Life?" Montreal Gazette 19 Apr. 1994.

Fuchs, Alfred H., and Katharine S. Milar. "Psychology as a science." Handbook of psychology (2003).

Gazzaniga, Michael S. "Forty-five years of split-brain research and still going strong." Nature Reviews Neuroscience 6.8 (2005): 653-659.

Gladwell, Malcolm. Blink: The power of thinking without thinking. Hachette Digital, Inc., 2007.

Hergenhahn, Baldwin, and Tracy Henley. An introduction to the history of psychology. Cengage Learning, 2013.

Hitchens, Christopher. Hitch-22: A memoir. Random House LLC, 2011.

Landois, Leonard. Textbook of human physiology. 1889.

Mueller, J. H., et al. "Research ethics: A tool for harassment in the academic workplace." Workplace mobbing in academe: Reports from 20 (2004): 290-313.

Müller, Joh. Elements of physiology. Lea and Blanchard, 1843.

Nebes, R. D., and R. W. Sperry. "Hemispheric deconnection syndrome with cerebral birth injury in the dominant arm area." Neuropsychologia 9.3 (1971): 247-259.

Poffenberger, Albert Theodor. Reaction time to retinal stimulation: with special reference to the time lost in conduction through nerve centers. No. 23. The Science press, 1912.

Sergent, Justine. "A new look at the human split brain." Brain 110.5 (1987): 1375-1392.

Sloan, Phillip R. "Descartes, the sceptics, and the rejection of vitalism in seventeenth-century physiology." Studies in History and Philosophy of Science Part A 8.1 (1977): 1-28.

Sperry, Roger W. "Chemoaffinity in the orderly growth of nerve fiber patterns and connections." Proceedings of the National Academy of Sciences of the United States of America 50.4 (1963): 703.

Zaidel, Eran, Dahlia W. Zaidel, and Joseph E. Bogen. "The split brain." Encyclopedia of neuroscience. Amsterdam: Elsevier Science (1999).

CHAPTER 6

Bever, Thomas G., and Robert J. Chiarello. "Cerebral dominance in musicians and nonmusicians." Science 185.4150 (1974): 537-539.

Bloom, Juliana S., and George W. Hynd. "The role of the corpus callosum in interhemispheric transfer of information: excitation or inhibition?." Neuropsychology review 15.2 (2005): 59-71.

Bolger, Donald J., Charles A. Perfetti, and Walter Schneider. "Cross-cultural effect on the brain revisited: Universal structures plus writing system variation." Human brain mapping 25.1 (2005): 92-104.

Champagne-Lavau, Maud, Emmanuel Stip, and Yves Joanette. "Language functions in right-hemisphere damage and schizophrenia: apparently similar pragmatic deficits may hide profound differences." Brain 130.2 (2007): e67-e67.

Cook, Norman D. "The transmission of information in natural systems." Journal of theoretical biology 108.3 (1984): 349-367.

David, Jack H. "The Two Sides of Music." Austin Community College. N.p., n.d. Web. 31 Mar. 2014.

Feinberg, Todd E., et al. "Two alien hand syndromes." Neurology 42.1 (1992): 19-19.

Freud, Sigmund. The future of an illusion. Broadview Press, 2012.

Gazzaniga, Michael S. "Cerebral specialization and inter-hemispheric communication Does the corpus callosum enable the human condition?." Brain 123.7 (2000): 1293-1326.

Gilbert, Daniel T. "How mental systems believe." American psychologist 46.2 (1991): 107.

Henik, Avishai, and Ruth Salo. "Schizophrenia and the stroop effect." Behavioral and cognitive neuroscience reviews 3.1 (2004): 42-59.

James, William. Psychology, briefer course. Vol. 14. Harvard University Press, 1984.

Jaynes, Julian. The origin of consciousness in the breakdown of the bicameral mind. Houghton Mifflin Harcourt, 2000.

Kahneman, Daniel. Thinking, fast and slow. Macmillan, 2011.

MacLeod, Colin M. "Half a century of research on the Stroop effect: an integrative review." Psychological bulletin 109.2 (1991): 163.

McGilchrist, Iain. The master and his emissary: The divided brain and the making of the western world. Yale University Press, 2009.

Ramachandran, V. S. "Split brain with one half atheist and one half theist." Youtube.

Ramachandran, V. S. and Sandra Blakeslee. Phantoms of the brain: Probing the mysteries of the human mind. William Morrow and Comp., 1998.

Reuter-Lorenz, P. A., et al. "Fate of neglected targets: a chronometric analysis of redundant target effects in the bisected brain." Journal of Experimental Psychology: Human Perception and Performance 21.2 (1995): 211.

Sala, Clelia Marchetti Sergio Della. "Disentangling the alien and anarchic hand." Cognitive neuropsychiatry 3.3 (1998): 191-207.

Sone, Toshimasa, et al. "Sense of life worth living (ikigai) and mortality in Japan: Ohsaki Study." Psychosomatic Medicine 70.6 (2008): 709-715.

Stroop, J. Ridley. "Studies of interference in serial verbal reactions." Journal of experimental psychology 18.6 (1935): 643.

Taleb, Nassim Nicholas. The Black Swan: The Impact of the Highly Improbable Fragility. Random House LLC, 2010.

Taylor, Jill. "Stroke of Insight." Youtube.

Zaidel, Eran, and Marco Iacoboni, eds. The parallel brain: The cognitive neuroscience of the corpus callosum. MIT press, 2003.

CHAPTER 7

A Hole in the Head. Dir. Eli Kabillio. Spectacle Films, 1998. Film.

Assaf, Yaniv, and Ofer Pasternak. "Diffusion tensor imaging (DTI)-based white matter mapping in brain research: a review." Journal of Molecular Neuroscience 34.1 (2008): 51-61.

Blos, V. T. "Cranial surgery in ancient Mesoamerica." (2003).

Blum, Alan. "A bedside conversation with Wilder Penfield." Canadian Medical Association Journal 183.7 (2011): 745-746.

Clower, William T., and Stanley Finger. "Discovering trepanation: the contribution of Paul Broca." Neurosurgery 49.6 (2001): 1417-1426.

Crania Americana, Or, A Comparative View of the Skulls of Various Aboriginal Nations of North and South America: To which is Prefixed an Essay on the Varieties of the Human Species. Philadelphia: J. Dobson; London: Simpkin, Marshall, 1983.

Galen, and Charles Singer. Galen on Anatomical Procedures: Translation of the Surviving Books; with

Introduction and Notes. Wellcome Historical Medical-Museum, 1956.

Herculano-Houzel, Suzana. "The human brain in numbers: a linearly scaled-up primate brain." Frontiers in human neuroscience 3 (2009): 31.

Mai, François. "Wilder Penfield, Man of Letters." The Canadian Journal of Neurological Sciences 39.6 (2012): 845-846.

Minsky, Marvin. "The society of mind." The Personalist Forum. 1987.

Missios, Symeon. "Hippocrates, Galen, and the uses of trepanation in the ancient classical world." Neurosurg Focus 23.1 (2007): E11.

Oakley, Kenneth Page, et al. Contributions on trepanning or trephination in ancient and modern times. Royal Anthropological Institute of Great Britain and Ireland, 1959.

Pearce, J. M. S. "Marie-Jean-Pierre Flourens (1794–1867) and Cortical Localization." European neurology 61.5 (2009): 311-314.

Penfield, Wilder, and Edwin Boldrey. "Somatic motor and sensory representation in the cerebral cortex of man as studied by electrical stimulation." Brain: A journal of neurology (1937).

Schwartz, Jeffrey M., and Sharon Begley. The mind and the brain: Neuroplasticity and the power of mental force. Regan Books/Harper Collins Publishers, 2002.

Vesalius, Andreas. De humani corporis fabrica libri septem. apud Franciscum Franciscium Senensem, & Ioannem Criegher Germanum, 1964.

Index

About the Author

Matt Gaidica is currently a student in the Neuroscience Graduate Program at the University of Michigan. He has a Bachelor's of Science degree in electrical engineering from Kettering University, and has worked in both Michigan and California in research and industry. Gaidica is currently investigating the role of deep brain circuits in health and disease. *Left, A History of the Hemispheres* is his first book.

www.ingramcontent.com/pod-product-compliance
Lightning Source LLC
Chambersburg PA
CBHW021404170526
45164CB00002B/499